Interactive Reader and Study

HOLT

California Social Studies
United States History
Independence to 1914

HOLT, RINEHART AND WINSTON

A Harcourt Education Company

Orlando • **Austin** • New York • San Diego • London

Contents

Contents

Chapter 1 Early Exploration and Settlement

HISTORY–SOCIAL SCIENCE STANDARDS

HSS 7.7 Students compare and contrast the geographic, political, economic, religious, and social structures of the Mesoamerican and Andean civilizations.

HSS 7.11 Students analyze political and economic change in the sixteenth, seventeenth, and eighteenth centuries (the Age of Exploration, the Enlightenment, and the Age of Reason).

HSS Analysis Skill

CHAPTER SUMMARY

CAUSE	EFFECT
Economic growth in Europe	Search for new trade routes to _____ and _____
Search for trade routes	Discovery of _____
Spanish conquest in Americas	Large empire and _____
Religious conflicts between Catholics and Protestants	Civil wars and war between _____ and _____

COMPREHENSION AND CRITICAL THINKING

1. Identify Cause and Effect To which continents did Europeans seek new trade routes in the 1400s? What was discovered as a result?

_____

_____

dentify Cause and Effect How did the Spanish conquests help Spain grow rich?

_____

_____

e Inferences How do you think religious conflicts affected Spanish colonies?

_____

_____

_____

Chapter 1 Early Exploration and Settlement

Section 1

MAIN IDEAS

1. Climate changes allowed people to migrate to the Americas.
2. Early societies existed in Mesoamerica and North America.
3. The environment influenced Native American cultures in North America.

 HSS 7.7
Students compare and contrast the geographic, political, economic, religious, and social structures of the Mesoamerican and Andean Civilizations.

Key Terms and People

Paleo-Indians the first Americans

migration a movement of people from one region to another

hunter-gatherers people who lived by hunting animals and gathering wild plants

environments climates and landscapes that surround living things

societies groups that share a common culture

culture a set of common values and traditions

totems ancestor or animal spirits

Iroquois League a political confederation made up of the Cayuga, Mohawk, Oneida, Onondaga, and Seneca peoples

Academic Vocabulary

method a way of doing something

develop grow or improve

Section Summary

THE FIRST MIGRATION TO THE AMERICAS

During the Ice Age, large amounts of water were frozen in huge glaciers. Ocean levels dropped, and a land bridge appeared between northeastern Asia and present-day Alaska. It is called the Bering Land Bridge. Artifacts show that **Paleo-Indians** moved into Alaska in a **migration** between 38,000 BC and 10,000 BC. Paleo-Indians were **hunter-gatherers**. The Ice Age ended about 8000 BC. Glaciers melted, so ocean levels rose. Water covered the Bering Land Bridge. **Environments** grew warmer. Herds of buffalo and deer flourished and were hunted by

Why did the Bering appear?

_____

2.

3. Mak

_____

Copyright © by Ho
Chapter 1

Native Americans. As environments changed, Native Americans began to domesticate wild plants and animals. Changing environments led to changing **societies**. Societies shared common **cultures**.

MESOAMERICAN AND NORTH AMERICAN SOCIETIES

The Olmec thrived on the Gulf Coast of Mexico from 1200 to 400 BC. They devised number and calendar systems and a method of writing. The Maya flourished in southeastern Mesoamerica from AD 300 to 900. They built large stone palaces and temples. Aztecs invaded the central valley of Mexico about 1200, conquering their neighbors to form an empire. North America had several farming cultures.

> **What systems did the Olmec devise?**
> _____
> _____

CULTURAL AREAS IN NORTH AMERICA

Culture areas, or geographic locations that influence societies, are used to describe ancient Native American peoples. The Aleut and Inuit of the Arctic fished and hunted large mammals. The Subarctic peoples hunted the caribou. Peoples of the Northwest carved **totems**. Most Native Americans in the California region lived in small groups. Native Americans in the dry West and Southwest of present-day United States fished for salmon if they lived near rivers. Many moved often in search of food. The Pueblo irrigated land to grow corn. Native Americans on the Great Plains hunted buffalo and other game that grazed in herds. In the East, the Algonquian and Iroquois peoples hunted and gathered plants to survive. The Iroquois put together the **Iroquois League**. The League waged war and made peace with non-Iroquois peoples. Women chose the male members of the League council and could overrule council decisions.

> **Why would Native Americans in the West and Southwest move often?**
> _____
> _____

> **What powers did Iroquois League women have?**
> _____
> _____

CHALLENGE ACTIVITY

Critical Thinking: Drawing Inferences Imagine you are a Paleo-Indian who is crossing the Bering Land Bridge into Alaska. Write a journal entry that describes how you feel. **HSS Analysis Skills HI 1, HI 4**

Chapter 1 Early Exploration and Settlement

MAIN IDEAS

1. Economic changes in Europe led to the discovery of new trade routes.
2. Christopher Columbus and other explorers discovered new continents.
3. The Columbian Exchange affected the Americas and Europe.

 HSS 7.11

Students analyze political and economic change in the sixteenth, seventeenth, and eighteenth centuries (the Age of Exploration, the Enlightenment, and the Age of Reason).

Key Terms and People

capital money or property that is used to earn more money

joint-stock companies businesses in which a group of people invests together

Christopher Columbus a sailor from Genoa, Italy

Ferdinand Magellan a Portuguese sea captain

Northwest Passage a path through North America that would allow ships to sail from the Atlantic to the Pacific

Columbian Exchange the transfer of plants, animals, and diseases from the "Old World" of Asia, Africa, and Europe to the "New World" of the Americas and from the Americas to Asia, Africa, and Europe

Section Summary

THE EUROPEAN ECONOMY GROWS

In Europe in the 1200s, the Commercial Revolution had changed the ways people carried on business.

Then, in the 1300s, Europeans experienced an epidemic called the Black Death, which killed millions of people. The European economy recovered. People raised or lowered prices to earn more. Farmers rented land to raise crops to sell for profit. Venice and other cities became rich trading centers. A class of wealthy people sprang up. Merchant families wanted **capital**. Merchants founded **joint-stock companies** that raised money while lowering personal risk.

What was an advantage of joint-stock companies?

_____

_____

TRADE WITH AFRICA AND ASIA

The biggest profits rose from trade with Africa and Asia. Overland trade made goods expensive. European merchants sought sea routes to Africa and

Underline the sentence that explains why Europeans wanted to cross the Atlantic.

Asia. Europeans wanted to cross the Atlantic to discover new trade routes.

PORTUGUESE EXPLORATIONS

Portugal led exploration in the early 1400s. Explorer Vasco da Gama left Lisbon and journeyed around the Cape of Good Hope, arriving in India and establishing the first trade route to that country.

CHRISTOPHER COLUMBUS

Christopher Columbus persuaded the Spanish king and queen to pay for his three ships to travel to North America. His ships landed in the Bahamas and found the Taino. He called them Indians.

OTHER EXPLORERS SET SAIL

Other European explorers also landed in North America. Amerigo Vespucci reached present-day South America, and a mapmaker labeled the continents across the ocean America in his honor. **Ferdinand Magellan** found the western route to Asia. Europeans realized Columbus had not discovered a sea route to Asia.

> **Why was America given its name?**
> _____
> _____

THE SEARCH FOR A NORTHWEST PASSAGE

European nations looked to North America to find a **Northwest Passage**. None was found, but explorations led to more interest in North America.

THE COLUMBIAN EXCHANGE

The **Columbian Exchange** made a huge difference in the world. Many of the changes were good. Other results, such as introducing deadly diseases to Native Americans, were bad.

> **Name one bad result from the Columbian Exchange.**
> _____
> _____

CHALLENGE ACTIVITY

Critical Thinking: Elaborating You are the sailor on Columbus's first voyage who first sighted land after the long trip. Write a short poem about the moment you saw land. What emotions were you feeling?
HSS Analysis Skills HR 2, HI 1, HI 3

Chapter 1 Early Exploration and Settlement

MAIN IDEAS

1. Spanish armies explored and conquered much of the Americas.

2. Spain used a variety of ways to govern its empire in the Americas.

 HSS 7.7
Students compare and contrast the geographic, political, economic, religious, and social structures of the Mesoamerican and Andean civilizations.

Key Terms and People

conquistadors Spanish soldiers who led military expeditions in the Americas

Hernán Cortés a conquistador sent by the governor of Cuba to present-day Mexico

Moctezuma II king of a wealthy land in present-day Mexico

Francisco Pizarro a conquistador who landed with a small army on the coast of present-day Peru

Junípero Serra Spanish missionary who traveled to California to spread Christianity and founded San Francisco and eight other missions

encomienda **system** a system that gave settlers the right to tax local American Indians or to make them work

Bartolomé de Las Casas a priest who defended American Indians' rights

plantations large farms that grew just one kind of crop and made huge profits for their owners

Section Summary

SPANISH EXPLORATION OF THE AMERICAS

Hernán Cortés led **conquistadors** to present-day Mexico in 1519. He heard of land to the west ruled by **Moctezuma II**, king of the Aztec Empire. The Aztec ruled several million people and had developed a rich civilization. Cortés had only 500 soldiers, a few horses and guns compared with thousands of Aztec warriors. Aztecs had never seen horses and were afraid. Cortés's weapons also gave him an advantage. Cortés imprisoned Moctezuma, although the king was friendly with the Spanish. The Aztec revolted. Thousands of troops from other American Indian groups helped Cortés conquer the Aztecs, as did diseases that the Spanish brought to Mexico. **Francisco Pizarro** heard talk of golden cities in South America's

Why might Cortés have wanted to conquer the Aztec?

_____

_____

What advantage did the Spanish have over the Aztec?

_____

_____

Why did Pizarro first enter the Inca Empire?

_____

_____

Andes mountains. With his army, he marched into the Inca Empire and kidnapped the Inca ruler. The Inca paid a ransom of gold and silver. Pizarro killed their ruler anyway. He allied himself with rebel Incas and conquered the empire by 1537. Spanish explorers claimed other areas too, including Florida and the Pacific coast of North America.

THE SPANISH EMPIRE

Spain became rich from its American colonies. From 1503 to 1660, Spanish fleets transported 200 tons of gold and 18,600 tons of silver from the American empire to Spain. Food for the Spanish empire came from Peru and Mexico. A system of royal officials enabled Spain to rule its huge empire. The Spanish had three types of settlements in New Spain. These played economic, religious, or military roles. Pueblos were trading posts and sometimes government centers. Priests started missions to convert American Indians to Catholicism. The Spanish constructed presidios, or military bases. Most Spanish colonists lived where they could gain wealth. Few lived in the borderlands. To link its communities, Spain built "the Royal Road," or El Camino Real. Among the last borderlands settled by the Spanish was California. Spanish missionary **Junípero Serra** founded San Francisco and eight other missions along the Pacific coast. To pay back settlers for their work, Spain established the *encomienda* **system**. The priest **Bartolomé de Las Casas** spoke out against this system. Many American Indians died of disease and exhaustion, so the Spanish began bringing enslaved Africans to work on **plantations** in New Spain.

> Underline the sentence that explains the roles that Spain's different kinds of settlements played.

> Why might Las Casas have spoken out against the *encomienda* system?
> _____
> _____

CHALLENGE ACTIVITY

Critical Thinking: Identifying Cause and Effect Make a chart showing the causes and effects of Spain's conquests in the Americas.

HSS Analysis Skills CS 2, HI 1

Chapter 1 Early Exploration and Settlement

MAIN IDEAS

1. Religious and political events in Europe affected the exploration of the Americas.
2. War between England and Spain affected settlement of North America.
3. The French, Dutch, and Swedish also tried to settle North America.

 HSS 7.11

Students analyze political and economic change in the sixteenth, seventeenth, and eighteenth centuries (the Age of Exploration, the Enlightenment, and the Age of Reason).

Key Terms and People

Protestant Reformation a religious movement that began as an effort to reform the Catholic Church

Protestants the reformers who protested the Catholic Church's practices

printing press an invention that helped spread the ideas of the Reformation

Spanish Armada a huge fleet of about 130 ships and 30,000 sailors

inflation a rise in the amount of money in use and in the price of goods

charter a document giving permission to start a colony

Section Summary

THE PROTESTANT REFORMATION

In 1517 a priest named Martin Luther launched the **Protestant Reformation**. His followers were called **Protestants**. Luther said the Catholic Church was too rich and abused its powers. The **printing press** helped spread Protestant ideas because large numbers of Bibles could be printed. More people could read the Bible on their own instead of depending on priests to explain it. Often conflicts between Catholics and Protestants caused civil war in Europe. In the late 1500s French Catholics fought French Protestants, known as Huguenots. Many Huguenots traveled to the Americas for religious freedom. In 1534 King Henry VIII established the Church of England, or Anglican Church. This was a Protestant church that still had many Catholic ceremonies, such as mass. Henry declared himself head of the church.

> How did the printing press help spread Protestant ideas?
> _____
> _____

> What were French Protestants called?
> _____
> _____

CONFLICT BETWEEN SPAIN AND ENGLAND

King Philip II of Spain, a Catholic ruler, put together a large fleet called the **Spanish Armada** to defeat the Protestant nation of England. England had fewer than 40 ships, but they were small and quick. In July, 1588, the English navy defeated the Armada. The defeat surprised the Spanish, whose economy was in trouble because of **inflation**.

> What helped England's navy defeat the Spanish Armada?
>
> _____

THE FRENCH EMPIRE

In the late 1600s the French branched out from the Great Lakes region. Louis Joliet and missionary Jacques Marquette reached the Mississippi. René-Robert de La Salle followed the river to the Gulf of Mexico and claimed the territory for France. France built small outposts throughout New France, trading for fur and allying itself with Native Americans.

> Describe the relationship between France and Native Americans.
>
> _____
>
> _____
>
> _____

NEW NETHERLAND AND NEW SWEDEN

The Dutch claimed land between the Delaware and Hudson Rivers and founded the town of New Amsterdam on Manhattan Island. Swedish settlers started New Sweden along the Delaware River. Peter Stuyvesant, governor of New Netherland, conquered New Sweden in 1655.

ENGLISH SETTLEMENT

In the late 1500s Sir Walter Raleigh received a **charter** to found a colony in present-day Virginia. The first colonists did not stay, but he sent more colonists. Those colonists disappeared. No one knows what happened to them.

> Why might English colonists have disappeared?
>
> _____
>
> _____

CHALLENGE ACTIVITY

Critical Thinking: Summarizing Write a bulleted list summarizing the early settlements of present-day North America. **HSS Analysis Skills CS 2, HI 1**

Chapter 2 The English Colonies

HISTORY–SOCIAL SCIENCE STANDARDS

HSS 8.1 Students understand the major events preceding the founding of the nation and relate their significance to the development of American constitutional democracy.

HSS 8.2 Students analyze the political principles underlying the U.S. Constitution and compare the enumerated and implied powers of the federal government.

HSS Analysis Skill CS 2 Students construct various time lines of key events, people, and periods of the historical era they are studying.

CHAPTER SUMMARY

Southern Colonies	New England Colonies	Middle Colonies
Farming	_____ _____	Trade and staple crops
_____ _____	Religion linked to government	Religious tolerance
_____ _____	Difficult start, soon prospered	_____ _____

COMPREHENSION AND CRITICAL THINKING

Use the answers to the following questions to fill in the graphic organizer above.

1. Classify Which colonies had difficulty in the beginning but soon flourished? Which colonies prospered from the start?

_____

2. Classify Which colonies did not rely on trade? In which was farming important?

_____

3. Compare and Contrast Compare and contrast religion in the three regions.

_____

_____

_____

Chapter 2 The English Colonies

MAIN IDEAS

1. The settlement of Jamestown was the first permanent English settlement in America.

2. Daily life in Virginia was challenging to the colonists.

3. Religious freedom and economic opportunities were motives for founding other southern colonies, including Maryland, the Carolinas, and Georgia.

4. Farming and slavery were important to the economies of the southern colonies.

 HSS 8.1
Students understand the major events preceding the founding of the nation and relate their significance to the development of American constitutional democracy.

Key Terms and People

Jamestown an English settlement in Virginia founded in 1607

John Smith a colonist and leader of Jamestown

Pocahontas a Powhatan Indian who married Jamestown colonist John Rolfe

indentured servants colonists who reached America by working for free for other people who had paid for their journeys

Bacon's Rebellion an uprising led by Nathaniel Bacon against high taxes

Toleration Act of 1649 an act that made limiting the religious rights of Christians a crime

Olaudah Equiano a former slave who wrote down his experiences

slave codes laws to control slaves

Academic Vocabulary

authority power, right to rule

Section Summary

SETTLEMENT IN JAMESTOWN

Life in **Jamestown** was hard. Few colonists knew how to grow crops for food. Captain **John Smith** worried about this. Many colonists starved. The Powhatan helped the colonists learn to grow crops.

    Pocahontas helped unite the Powhatan and the colonists, but she died in 1617. Fighting broke out between the colonists and the Powhatan and went

Why did many colonists in Jamestown starve?

_____

_____

on for the next 20 years. The colony existed under
the **authority** of a governor chosen by the king.

DAILY LIFE IN VIRGINIA

Colonists began forming large farms called plantations. At first **indentured servants** worked on plantations. In 1619 the first Africans came to Virginia.
Wealthy farmers began to use slave labor.

In 1676 Nathaniel Bacon, a wealthy frontier farmer, led **Bacon's Rebellion**. Bacon and his followers
burned Jamestown.

> What happened to Jamestown in 1676?
>
> _____
>
> _____

OTHER SOUTHERN COLONIES

Maryland was founded south of Virginia as a new
colony for Catholics. In the 1640s, Protestants
began moving there. Religious problems divided
Protestants and Catholics. The **Toleration Act of
1649** made limiting religious rights of Christians a
crime in Maryland.

> Circle the sentence that explains what the Toleration Act of 1649 did.

The Carolinas and Georgia were formed south of
Virginia and Maryland. South Carolina had many
large plantations, and owners bought slaves to work
on them. In Georgia many huge rice plantations
were worked by thousands of slaves.

ECONOMIES OF THE SOUTHERN COLONIES

The economies of the southern colonies were
based on farming. Many small farms and some
small plantations meant a large group of workers
was needed. African slaves became these workers.
Slavery was brutal. A former slave named **Olaudah
Equiano** wrote that slaves were often tortured,
murdered, and treated with barbarity. Most of the
southern states passed **slave codes** to control slaves.

> How did a former slave describe treatment of slaves?
>
> _____
>
> _____

CHALLENGE ACTIVITY

Critical Thinking: Designing Design a time line showing the dates of
important events in the colonies. **HSS Analysis Skills CS 1, CS 2, HI 1**

Section 2

MAIN IDEAS

1. The Pilgrims and Puritans came to America to avoid religious persecution.

2. Religion and education were closely linked in the New England colonies.

3. The New England economy was based on trade and farming.

4. Education was important in the New England colonies.

 HSS 8.1

Students understand the major events preceding the founding of the nation and relate their significance to the development of American constitutional democracy.

Key Terms and People

Puritans a Protestant group that wanted to reform, or purify, the Church of England

Pilgrims a Protestant group that cut all ties with the Church of England and was punished

immigrants people who have left the country of their birth to live in another country

Mayflower Compact a legal contract male passengers on the *Mayflower* signed agreeing to have fair laws to protect the general good

Squanto a Patuxet Indian who had lived in Europe and spoke English

John Winthrop the leader of Puritans who left England for Massachusetts seeking religious freedom

Anne Hutchinson a Puritan who claimed to receive her religious views directly from God and who was forced to leave the Massachusetts Bay Colony

Section Summary

PILGRIMS AND PURITANS

The **Pilgrims** were a group of **Puritans** who suffered persecution in England. They became **immigrants**, first settling in the Netherlands and then sailing to America. When they reached America, they signed the **Mayflower Compact**. This was one of the first times English colonists tried to govern themselves. Earlier, in 1215, English nobles had forced the king to give them some rights in the Magna Carta. Later the English Bill of Rights provided more liberties.

The Pilgrims learned to fertilize their soil from **Squanto**. They invited him and 90 Wampanoag guests to a feast now known as Thanksgiving.

> Name two early examples of the English receiving rights.
>
> _____
>
> _____

Religion and education played important roles in the Pilgrims' lives, which centered on families. Everyone worked hard. Women had rights that they did not have in England.

Puritans and merchants founded the Massachusetts Bay colony. Tens of thousands of English men, women and children would immigrate to it. **John Winthrop** led one group. Puritans believed they had a sacred agreement with God to build a Christian colony.

> What was the Puritans' sacred agreement with God?
>
> _____
>
> _____

RELIGION AND GOVERNMENT IN NEW ENGLAND

Politics and religion were closely linked in Puritan New England. Some self-government existed. However, only the chosen male church members could vote.

> Underline the sentence that means women could not vote in Puritan New England.

Some Puritans had different religious views than others. Minister Roger Williams supported the separation of the church from politics. He founded Providence. **Anne Hutchinson** was forced to leave the colony because of her religious ideas.

NEW ENGLAND ECONOMY

The New England colonies had a hard climate and rocky soil. The kind of farming done in Virginia was impossible there. Instead, they traded goods, fished, built ships, and became skilled craftspeople.

> Compare sources of income in Virginia and New England.
>
> _____
>
> _____

EDUCATION IN THE COLONIES

New England parents wanted their children to read the Bible. They made laws requiring the education of children. The colonists also founded Harvard College to teach ministers.

CHALLENGE ACTIVITY

Critical Thinking: Developing Questions Develop three questions about the Pilgrims' contributions and research to answer them. **HSS Analysis Skills HR 1, HI 2**

Chapter 2 The English Colonies

MAIN IDEAS

1. The English created New York and New Jersey from former Dutch territory.

2. William Penn established the colony of Pennsylvania.

3. The economy of the middle colonies was supported by trade and staple crops.

 HSS 8.1
Students understand the major events preceding the founding of the nation and relate their significance to the development of American constitutional democracy.

Key Terms and People

Peter Stuyvesant director general who took control of New Amsterdam beginning in 1647

Quakers a Protestant religious group founded by George Fox in the mid-1600s in England

William Penn a Quaker leader who began the Pennsylvania colony

staple crops crops that are always needed, such as wheat, barley, and oats

Section Summary

NEW YORK AND NEW JERSEY

In 1613 the Dutch formed New Netherland as a base for trading fur with the Iroquois. They traded fur mostly in the town of New Amsterdam on Manhattan Island. Large land grants and religious tolerance meant Jews, French Huguenots, Puritans, and others came to the colony. **Peter Stuyvesant** ruled the colony for many years. Then in 1664 an English fleet gained control of New Netherland without any fighting. New Amsterdam became New York City, named in honor of the Duke of York. New York was the first of the middle colonies.

The Duke of York made two men proprietors, or governors, of New Jersey. The colony rested between the Hudson and Delaware Rivers. Dutch, Finns, Swedes, Scots, and others lived there.

> What was the first name of New York City?
> _____
> _____

> Why did the Dutch settle New Amsterdam?
> _____
> _____

> Underline the sentence that makes you think the population of New Jersey was diverse.

PENN'S COLONY

One of the biggest religious groups in New Jersey was the Society of Friends, or the **Quakers**. Their religious practices were different. They believed in the equality of men and women before God. They also backed religious tolerance for all groups. The Quakers' beliefs angered many. They were treated badly in both England and America.

William Penn started a colony named Pennsylvania. He offered religious freedom to all Christians. He created a way to change colony laws based on what the people wanted. Many Quakers settled in Pennsylvania. Penn named his capital Philadelphia, which means "the city of Brotherly Love."

> Why was Pennsylvania's capital named Philadelphia?
> _____
> _____

ECONOMY OF THE MIDDLE COLONIES

A good climate and fertile land meant the colonists could grow a large quantity of **staple crops**, unlike colonists in New England. Some slaves worked in the middle colonies, but not as many as in the south. Indentured servants did more of the labor.

By the 1700s Philadelphia and New York City had grown into large cities. Trade was important to the middle colonies. Women ran some businesses and practiced as doctors, nurses, or midwives.

> How did the middle and southern colonies differ?
> _____
> _____

CHALLENGE ACTIVITY

Critical Thinking: Evaluating Think about the two colonies. How are they similar? How are they different? Decide which colony you would like to live in, then write a short essay explaining why you chose the colony you did. Illustrate your essay. **HSS Analysis Skills HR 3, HI 2**

Chapter 2 The English Colonies

Section 4

MAIN IDEAS

1. Colonial governments were influenced by political changes in England.

2. English trade laws limited free trade in the colonies.

3. The Great Awakening and the Enlightenment led to ideas of political equality among many colonists.

4. The French and Indian War gave England control of more land in North America.

 HSS 8.1

Students understand the major events preceding the founding of the nation and relate their significance to the development of American constitutional democracy.

Key Terms and People

town meeting colonists decided issues and made laws in town meetings

English Bill of Rights an act that reduced the powers of the English monarch and gave Parliament more power in 1689

triangular trade indirect trade between the American colonies and Britain

Middle Passage name for the slaves' voyage across the Atlantic

Great Awakening an awakening in the religious lives of colonists

Enlightenment its thinkers used reason and logic, as scientists had done

Pontiac American Indian leader who led a rebellion in the Ohio Valley in 1763

Section Summary

COLONIAL GOVERNMENTS

The House of Burgesses helped make laws in Virginia. In New England colonists at the **town meeting** decided local issues. The middle colonies used both county courts and town meetings.

> **How were laws made in Virginia and New England?**
> _____
> _____

King James II of England thought the colonies were too independent. He united the northern colonies and limited their powers. In 1689 the **English Bill of Rights** shifted power from the monarch to Parliament, the British governing body. These rights were not extended to the colonists.

> **Did the colonists benefit from the English Bill of Rights?**
> _____

ENGLISH TRADE LAWS

England controlled its American colonies partly to earn money. Parliament passed Navigation Acts that required colonists to trade only with Britain. However, some colonists wanted to buy and sell goods at the market offering the best prices.

In a deadly version of **triangular trade**, New England colonists traded rum for slaves from the African coast. The slave trade brought 10 million Africans across the Atlantic Ocean. In the **Middle Passage** thousands of them died.

> What is the name given to the voyage of slaves from Africa to America?
>
> _____

GREAT AWAKENING AND ENLIGHTENMENT

During the **Great Awakening** talk of the spiritual equality of all people made some people think about political equality. John Locke, an **Enlightenment** thinker, said people should obey their rulers only if the state protected life, liberty, and property.

In 1675 a war erupted between New England colonists and some American Indians. Metacomet, who was also known as King Philip, led the Wampanoag. Each side killed men, women, and children from the other. The fighting ended in 1676.

> What was King Philip's real name?
>
> _____

THE FRENCH AND INDIAN WAR

The British and the French both wanted to control certain territory in North America. The **French and Indian War** was about the British wanting to settle in the Ohio Valley and the French wanting it for the fur trade. After the war, Britain received Canada and all French lands east of the Mississippi River.

The Ohio Valley proved good for farming, but Indian leaders opposed British settlements. American Indian Chief **Pontiac** led followers against the British. He later gave up, but King George III banned colonists from settling on Indian lands. Many settlers ignored the ban.

> How did many Americans react to the king's ban on settling on Indian lands?
>
> _____

CHALLENGE ACTIVITY

Critical Thinking: Drawing Inferences Imagine you live during the Enlightenment. Write a short journal entry describing the time.
HSS Analysis Skills HR 3, HI 1

Chapter 2 The English Colonies

MAIN IDEAS

1. British efforts to raise taxes on colonists sparked protest.

2. The Boston Massacre caused colonial resentment toward Great Britain.

3. Colonists protested the British tax on tea with the Boston Tea Party.

4. Great Britain responded to colonial actions by passing the Intolerable Acts.

 HSS 8.1
Students understand the major events preceding the founding of the nation and relate their significance to the development of American constitutional democracy.

Key Terms and People

Samuel Adams Boston leader who believed Parliament could not tax the colonists without their permission

Stamp Act of 1765 required colonists to pay for an official stamp when buying paper items

Boston Massacre shootings by British soldiers killed five colonists

Tea Act an act allowing a British company to sell cheap tea directly to the colonists

Boston Tea Party colonists dressed as American Indians dumped 340 tea chests from British ships into Boston Harbor

Intolerable Acts laws passed to punish colonists for the Boston Tea Party

Section Summary

GREAT BRITAIN RAISES TAXES

Parliament raised the colonists' taxes for money to pay for the French and Indian War and a British army kept in North America to protect the colonists against American Indian attacks. Parliament also tried harder to arrest smugglers avoiding taxes.

> **Name one reason that Parliament raised taxes.**
> _____
> _____

Many colonists believed Britain had no right to tax them without their permission. **Samuel Adams** and James Otis spread the slogan "No Taxation without Representation." Colonists chose to boycott, refusing to buy British goods. They hoped Parliament would end the new taxes. The **Stamp Act of 1765** meant a tax had to be paid on legal documents, licenses, and other items.

> **What is taxation without representation?**
> _____
> _____
> _____

Section 5, continued

The Townshend Acts charged taxes on imported glass, lead, paints, paper, and tea. Boston's Sons of Liberty attacked the customs houses to protest the taking of a ship on suspicion of smuggling. British soldiers came in 1768 to restore order.

> Underline the sentence that tells what the Townshend Acts did.

BOSTON MASSACRE

On March 5, 1770, a few troops fired on Bostonians throwing snowballs at them. That led to the **Boston Massacre**. The soldiers and their officer were charged with murder. A jury found the officer and six soldiers acted in self-defense and were not guilty. Two soldiers were convicted of accidental killing. This calmed Boston for a while.

> Why do you think the jury found some of the troops not guilty?
> _____
> _____
> _____

THE BOSTON TEA PARTY

Parliament ended almost all the Townshend Acts but left the tax on tea. Colonists united against the **Tea Act**. In November 1773 the **Boston Tea Party** showed the colonists' spirit of rebellion.

THE INTOLERABLE ACTS

The Boston Tea Party made the new British Prime Minister very angry. Parliament punished Boston by passing the **Intolerable Acts**. The laws closed Boston Harbor until the colonists paid for the lost tea. They also had other effects unacceptable to the colonists and angered the colonists even more.

> Why did the Boston Tea Party anger the British Prime Minister?
> _____
> _____
> _____

CHALLENGE ACTIVITY

Critical Thinking: Imagining Imagine you write for Boston's Committee of Correspondence and pen a brief description of the Boston Massacre.
HSS Analysis Skills CS 1, HR 2, HR 5

Chapter 3 The American Revolution

HISTORY–SOCIAL SCIENCE STANDARDS
HSS 8.1 Students understand the major events preceding the founding of the nation and relate their significance to the development of American constitutional democracy.
HSS Analysis Skill HI 5 Students recognize that interpretations of history are subject to change.

CHAPTER SUMMARY

Declaration of Rights	led to	"Shot heard around the world"
Declaration of Independence	led to	Break with Britain
Battle of Saratoga	led to	
Battle of Yorktown	led to	

COMPREHENSION AND CRITICAL THINKING

Use the answers to the following questions to fill in the graphic organizer above.

1. Recall Which battle was the turning point of the Revolutionary War?

_____

2. Identify Cause and Effect Which battle led to the victory of the Patriot forces in the Revolutionary War?

_____

3. Evaluate Do you think the Patriots could have won the Revolutionary War without help from other countries? Why or why not?

_____

_____

Chapter 3 The American Revolution

MAIN IDEAS

1. The First Continental Congress demanded certain rights from Great Britain.

2. Armed conflict between British soldiers and colonists broke out with the "shot heard 'round the world."

3. The Second Continental Congress created the Continental Army to fight the British.

 HSS 8.1
Students understand the major events preceding the founding of the nation and relate their significance to the development of American constitutional democracy.

Key Terms and People

First Continental Congress delegates from all the colonies except Georgia who met in Philadelphia in September 1774 to discuss how to respond to Britain

minutemen the members of the civilian volunteer militia

Redcoats British soldiers wearing red uniforms

Second Continental Congress delegates from 12 colonies who met in Philadelphia in May 1775

Continental Army army created by the second Congress to carry out the fight against Britain

George Washington the Virginian who commanded the Continental Army

Battle of Bunker Hill battle won by the British but with more than 1,000 casualties, about double the American losses

Academic Vocabulary

reaction response

Section Summary

FIRST CONTINENTAL CONGRESS

At the **First Continental Congress** delegates debated whether violence was certain or peace was necessary. They stopped all trade with Britain and told the colonial militias to prepare for war. They also drafted a Declaration of Rights. King George refused to consider it.

> **What did the delegates debate?**
> _____
> _____

"SHOT HEARD 'ROUND THE WORLD"

On April 19, 1775, 700 **Redcoats** set out for

Concord, where the colonists had a main weapons storehouse. A British general sent the soldiers to destroy it. Three colonists rode out on horseback to warn that the British were coming. Signals called out the **minutemen**. Seventy armed minutemen waited when the British reached Lexington. The British killed 8 minutemen, then went on to Concord where they destroyed a few weapons. As the British headed back to Boston, minutemen fired on them. More than 250 British soldiers were dead, wounded, or missing. The minutemen had fewer than 100 dead.

> Did the battle hurt the British or the colonists more?
> _____

SECOND CONTINENTAL CONGRESS

At the **Second Continental Congress,** some delegates said war would happen; others wanted peace. The Congress named the Massachusetts militia the **Continental Army**. The army's commander was **George Washington**. Delegates signed the Olive Branch Petition, asking King George to make peace. He would not consider it.

> What was King George's response to the Olive Branch Petition?
> _____

Meanwhile, minutemen kept the British inside Boston. On June 17, 1775, the British rose to find colonial forces dug in on Breed's Hill. The British force would have to cross the harbor in boats and fight its way up the hill. The British attacked. At last, the colonists opened fire. The British took the hill on their third try. But the **Battle of Bunker Hill** (named for a nearby hill) proved the colonists could hold their own against the British.

> Underline the sentence that best explains why the Battle of Bunker Hill was meaningful.

Two weeks later, General George Washington took command of the Continental Army of about 14,000 men in Boston. He set up cannons captured from Fort Ticonderoga to fire on the British from a hill out of the reach of British guns. On March 7, the British retreated from Boston.

CHALLENGE ACTIVITY

Critical Thinking: Analyzing List several rights we have as U.S. citizens.
HSS Analysis Skills HR 5, HI 2, HI 3

Chapter 3 The American Revolution

Section 2

MAIN IDEAS

1. Thomas Paine's *Common Sense* led many colonists to support independence.

2. Colonists had differing reactions to the Declaration of Independence.

 HSS 8.1
Students understand the major events preceding the founding of the nation and relate their significance to the development of American constitutional democracy.

Key Terms and People

Common Sense a 47-page pamphlet that argued against British rule over America

Thomas Paine author of *Common Sense*, who wrote that citizens, not monarchs, should make laws

Thomas Jefferson the main author of the Declaration of Independence

Declaration of Independence the document that formally announced the colonies' break from Great Britain

Patriots colonists who chose to fight for independence

Loyalists colonists, sometimes called Tories, who remained loyal to Britain

Section Summary

PAINE'S *COMMON SENSE*

Common Sense was published anonymously, or without the name of its author, who was **Thomas Paine**. At this time, the idea that citizens should pass laws made news. As word of the pamphlet spread throughout the colonies, it eventually sold about 500,000 copies. The pamphlet made a strong case for political and economic freedom. It supported the right to military self-defense. *Common Sense* changed the way many colonists viewed their king. In it Paine defied British oppression.

> Why do you think *Common Sense* was so popular?
>
> _____
>
> _____

INDEPENDENCE FOR COLONIES

The first point argued by **Thomas Jefferson** in The **Declaration of Independence** was that all men possess unalienable rights, or rights that cannot be denied. These rights include "life, liberty, and the pursuit of happiness." Jefferson also maintained

that King George III had trampled on the colonists' rights by supporting unfair laws and wrongly
meddling in colonial governments. In addition,
Jefferson argued that the colonies had the right
to independence from Britain. He believed in the
Enlightenment view of the social contract. This idea
says that citizens should agree to be governed only
when rulers and governments support their rights.
Jefferson said that King George III had violated the
social contract, so the colonies should not obey his
laws.

> **The Declaration of Independence is easy to find. Read it again.**

> **Why did Jefferson think the colonies should not obey King George III?**
>
> _____
>
> _____

On July 4, 1776, the Continental Congress voted
in favor of the Declaration of Independence. In
approving the Declaration, the Congress finally
broke away from Great Britain. Today we celebrate
the Fourth of July as the birthday of our nation.

Not everyone rejoiced over the approval of the
Declaration. **Patriots** and **Loyalists** became divided.
More than 50,000 Loyalists opposed the Revolution.
Sometimes family members were on opposite sides
during the war.

> **What did some families experience during the war?**
>
> _____
>
> _____

Looking back, we realize that the Declaration
paid no attention to many colonists. Abigail Adams,
wife of delegate John Adams, tried to influence him
to include women in the Declaration. It did not
happen. Enslaved African Americans also had no
rights under the Declaration. Slavery was legal in all
colonies in July 1776. The Revolutionary War would
not end the battle over slavery, even though New
England states moved to end it by 1780.

> **Name two groups who had no rights under the Declaration.**
>
> _____
>
> _____

CHALLENGE ACTIVITY

Critical Thinking: Hypothesizing You are a delegate to the Second
Continental Congress. Write and deliver a two-minute speech arguing
that the unalienable rights listed in the Declaration should also apply to
women and slaves. **HSS Analysis Skills HR 1, HI 1, HI 2, HI 3**

Chapter 3 The American Revolution

Section 3

MAIN IDEAS

1. Many Americans contributed to the war effort.

2. Despite early defeats to Britain, the Patriots claimed some victories.

3. Saratoga was a turning point in the war.

4. The winter at Valley Forge tested the strength of Patriot forces.

5. The war continued at sea and in the West.

 HSS 8.1
Students understand the major events preceding the founding of the nation and relate their significance to the development of American constitutional democracy.

Key Terms and People

mercenaries foreign soldiers who fight not out of loyalty, but for pay

Battle of Trenton a battle won by the Patriots against mercenary Hessians

Battle of Saratoga a great victory for the American forces in which British General John Burgoyne surrendered his entire army to American General Horatio Gates

Marquis de Lafayette a French nobleman who joined the Continental Army, then gave $200,000 of his money for the Revolution and influenced France to support it

Bernardo de Gálvez the governor of Spanish Louisiana, who became a Patriot ally

John Paul Jones a famous brave and clever marine commander

George Rogers Clark a commander who led Patriots to capture a British trading village and a fort in the West

Academic Vocabulary

strategy a plan for fighting a battle or war

Section Summary

AMERICANS AND THE WAR EFFORT

More than 230,000 mostly young soldiers served in the Continental Army. Some saw military life as an opportunity. Free African Americans served. Native Americans fought on both sides. A few women dressed as men and fought in battles.

> Underline the sentence that tells how some women chose to serve the war effort.

EARLY DEFEATS

George Washington had moved his troops to New York. However, the British force pushed the

Continental Army off of Long Island and into New
Jersey. The British General Howe, who thought the
rebellion would soon be over, left New Jersey in
the hands of Hessian **mercenaries** fighting for the
British. The Patriots surprised the Hessians and
won the **Battle of Trenton**.

> Name an early victory for the Patriots.

TURNING POINT AT SARATOGA

British General John Burgoyne planned to capture
the Hudson River valley. Burgoyne was outnumbered, and the Americans won their greatest victory so far at the **Battle of Saratoga**. Then Britain's
mighty foes, France and Spain, began to help the
Patriots. Holland also aided them. The **Marquis de
Lafayette** and **Bernardo de Gálvez** joined the war
on the Patriots' side.

> Underline the sentence about the success of the Patriots at the Battle of Saratoga.

WINTER AT VALLEY FORGE

The winter of 1777 turned brutally cold and snowy.
General Washington settled his troops at Valley
Forge, where they bore hardships with courage and
drilled to become better soldiers.

> What do you think the British thought about the war during the cold, quiet winter of 1777–78?
> _____
> _____

WAR AT SEA AND IN THE WEST

The small Continental Navy avoided the huge
British fleet, but sunk hundreds of individual
British ships. **John Paul Jones** fought a battle with
the British in which his ship took heavy damage. He
fought on, and the British ship surrendered. In the
West, **George Rogers Clark** led Patriots in taking a
British trading village and retaking a town. He also
convinced some Indian leaders to stay neutral.

> Why do you think the Continental Navy avoided the British naval fleet?
> _____
> _____

CHALLENGE ACTIVITY

Critical Thinking: Synthesizing Think about Continental Army soldiers
enduring hardships. Write and perform a dialogue between two soldiers
discussing their hard times at Valley Forge. **HSS Analysis Skills HR 2,
HI 1, HI 2, HI 4**

Chapter 3 The American Revolution

Section 4

> **MAIN IDEAS**
> 1. Patriot forces faced many problems in the war in the South.
> 2. The American Patriots finally defeated the British at the Battle of Yorktown.
> 3. The British and the Americans officially ended the war with the Treaty of Paris of 1783.

 HSS 8.1
Students understand the major events preceding the founding of the nation and relate their significance to the development of American constitutional democracy.

Key Terms and People

Francis Marion a Patriot leader who used hit-and-run attacks, known as guerilla warfare

Comte de Rochambeau commander of 4,000 French troops that aided the Patriot forces at the Battle of Yorktown

Battle of Yorktown the most important battle of the Revolutionary War, won by the Patriots

Treaty of Paris of 1783 the peace agreement in which Great Britain recognized the independence of the United States

Section Summary

WAR IN THE SOUTH

The war in the northern colonies did not go as the British government had hoped. The northern Patriots were tough to beat. The British moved the war into the South, where they believed large groups of Loyalists would help them win. General Henry Clinton led the British troops. The British plan worked at first. The war in the South proved especially nasty. Patriots and Loyalists engaged in direct fighting. The British wiped out crops, farm animals, and property. Georgia fell to the British. Next, the British conquered the port of Charleston, South Carolina. The Patriots failed to retake Camden, South Carolina. Patriot General Nathanael Greene arrived to shape up the army. He said he had never seen such a wasteland. Meanwhile, under

> Why did the British move the war to the South?
> _____
> _____

> Underline the sentence that explains how the British army waged war in the South.

the leadership of **Francis Marion**, Southern patriots used surprise attacks to cut off British supply and communication lines. The British could not capture Marion and his men.

> **How did Francis Marion and his men evade the British?**
> _____
> _____

BATTLE OF YORKTOWN

The Patriots were in trouble in early 1781. They had little money for paying soldiers and buying supplies. The British held most of the South as well as Philadelphia and New York. The Continental Army began to pressure the British in the Carolinas. General George Cornwallis moved his 7,200 men to Yorktown, Virginia. In New York, General Washington combined his troops with French troops commanded by **Comte de Rochambeau**. Washington marched his force to Virginia. With 16,000 soldiers, Washington's force surrounded Cornwallis. For weeks the French-American force wore down the British troops. Finally, the British surrendered. The Patriots captured 8,000 British prisoners at the **Battle of Yorktown**.

THE TREATY OF PARIS

Britain lost most of its army at Yorktown and could not fund a new one. So Great Britain and America began peace talks. Delegates took more than two years to reach a peace agreement. **The Treaty of Paris of 1783** gave the United States independence from Great Britain. It also created America's borders. In a separate treaty, Britain returned Florida to the Spanish. The Patriots' courage had won the Revolutionary War.

> **Why might reaching a peace treaty have taken so long?**
> _____
> _____

CHALLENGE ACTIVITY

Critical Thinking: Hypothesizing Imagine that the Patriots had lost the Revolutionary War. Help lead a class discussion on how your lives would be different today. **HSS Analysis Skills HI 1, HI 2, HI 3, HI 4**

Chapter 4 Forming a Government

HISTORY–SOCIAL SCIENCE STANDARDS
HSS 8.2 Students analyze the political principles underlying the U.S. Constitution and compare the enumerated and implied powers of the federal government.
HSS 8.3 Students understand the foundation of the American political system and the ways in which citizens participate in it.
HSS 8.9 Students analyze the early and steady attempts to abolish slavery and to realize the ideals of the Declaration of Independence.

CHAPTER SUMMARY

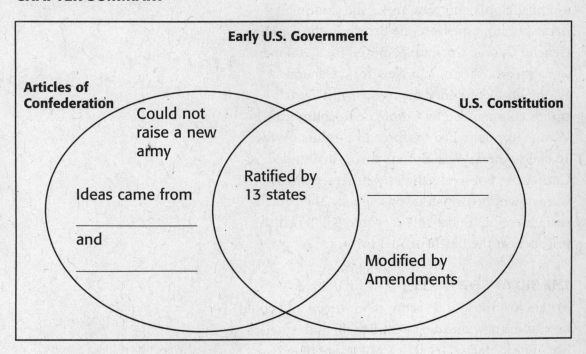

Early U.S. Government

Articles of Confederation

Could not raise a new army

Ideas came from _____ and _____

Ratified by 13 states

U.S. Constitution

_____

_____

Modified by Amendments

COMPREHENSION AND CRITICAL THINKING

Use the answers to the following questions to fill in the graphic organizer above.

1. Identify What sources gave Americans ideas for their first national government under the Articles of Confederation?

_____

_____

2. Compare and Contrast Describe four ways in which the U.S. Constitution was different from the Articles of Confederation.

_____

_____

Chapter 4 Forming a Government

MAIN IDEAS

1. The American people examined many ideas behind government.

2. The Articles of Confederation laid the base for the first government of the United States.

3. The Confederation Congress established the Northwest Territory.

 HSS 8.3

Students understand the foundation of the American political system and the ways in which citizens participate in it.

Key Terms and People

Magna Carta an English document that limited the power of the monarch

constitution a set of basic principles and laws that states the powers and duties of the government

English Bill of Rights the bill declared the power of Parliament and kept the monarch from passing new taxes or changing laws without Parliament's approval

Virginia Statute for Religious Freedom a law that included Thomas Jefferson's ideas granting religious freedom

suffrage voting rights

Articles of Confederation the new national constitution, which made a new Confederation Congress the national government

ratification official approval of the Articles by the states

Land Ordinance of 1785 a law that set up a system for surveying land and dividing the Northwest Territory

Northwest Ordinance of 1787 a law that established the Northwest Territory and formed a political system for the region

Northwest Territory a territory including Illinois, Indiana, Michigan, Ohio, and Wisconsin

Section Summary

IDEAS BEHIND GOVERNMENT

After winning independence from Great Britain, the United States needed to form new governments. The Americans first looked to English law for ideas. The **Magna Carta** and the **English Bill of Rights** gave them inspiration. Ideas from the Enlightenment also influenced them. English philosopher John Locke had thought the

Where did Americans find ideas for their government?

_____

_____

Section 1, continued

government had a duty to guard people's rights. In 1639 the people of Connecticut had created a government plan considered to be the first written **constitution** in the colonies. During the American Revolution nearly every state wrote a constitution to ensure that citizens elected representatives to make laws. **Suffrage** varied considerably from state to state. Some constitutions banned slavery. The **Virginia Statute for Religious Freedom** was an example of a law providing religious freedom.

> Do you think the states that banned slavery should have been more insistent that other states ban it also? Why or why not?
>
> _____
> _____
> _____

ARTICLES OF CONFEDERATION

The Continental Congress named a Committee of Thirteen, with one member from each colony. This committee drafted the **Articles of Confederation**. Under the Articles, the Confederation Congress had limited powers to guard the people's freedoms. Each state had one vote in the Congress. The Congress had powers, but it could only ask the states for money and soldiers. States could refuse these requests. After some conflicts, the Articles were **ratified** by all the states to form the first American government.

> Why would you have voted for or against the Articles?
>
> _____
> _____

NORTHWEST TERRITORY

Congress decided to raise money to pay debts by selling the ordinance lands. Congress passed the **Land Ordinance of 1785**. The **Northwest Ordinance of 1787** formed the **Northwest Territory**. It was then split up into several smaller territories. When the population of a territory hit 60,000, its settlers could draft their own constitution and ask to join the Union. Slavery was banned in the Northwest Territory.

> Underline the sentence that explains when a territory could ask to join the Union.

CHALLENGE ACTIVITY

Critical Thinking: Elaborating You plan to settle in the Northwest Territory. List 10 items you will take with you and explain why you chose the items you did. **HSS Analysis Skills HR 2, HI 1**

Chapter 4 Forming a Government

Section 2

MAIN IDEAS

1. The United States had difficulties with international relations.

2. Internal economic problems plagued the new nation.

3. Shays's Rebellion pointed out weaknesses in the Articles of Confederation.

4. Many Americans called for changes in the national government.

 HSS 8.2

Students analyze the political principles underlying the U.S. Constitution and compare the enumerated and implied powers of the federal government.

Key Terms and People

tariffs taxes on imports or exports

interstate commerce trade between two or more states

depression a period of low economic activity combined with a rise in unemployment

Daniel Shays a poor farmer and Revolutionary War veteran

Shays's Rebellion an uprising in which Daniel Shays led hundreds of men in a forced shutdown of the Supreme Court in Springfield, Massachusetts

Section Summary

INTERNATIONAL RELATIONS

The Continental Army broke up soon after the signing of the Treaty of Paris of 1783. The Articles of Confederation provided no way to raise a new army. The United States had a hard time guarding against foreign threats. Problems also arose in trading with Britain, which closed many British ports to U.S. ships. The British also forced American merchants to pay high **tariffs**. U.S. merchants increased prices to pay them, and costs were passed on to customers. In 1784 Spanish officials shut down the lower Mississippi River to U.S. shipping. Western farmers and merchants used the river to ship goods east and overseas. The U.S. government failed to work out an agreement with Spain. Critics thought Spain would have negotiated longer if America had a strong military force. The loss of the British West Indies

> **Point out one weakness in the Articles of Confederation.**
> _____
> _____

> **Why did tariffs hurt U.S. citizens?**
> _____
> _____

markets meant farmers could not sell goods there. U.S. exports dropped while lower-priced British goods kept entering America. Congress could not pass tariffs.

ECONOMIC PROBLEMS

Trade problems among the states, war debt, and a poor economy hurt the states. The Confederation Congress had no power to regulate **interstate commerce**. States looked out only for their own trade interests. In addition, states had trouble paying off war debts. They printed paper money, but it had no gold or silver backing and little value. This caused inflation, which occurs when increased prices for goods and services combine with the reduced value of money. The loss of trade with Britain coupled with inflation created a **depression**.

> Underline the sentence that lists problems facing the states.

> What conditions caused a depression in the United States?
> _____
> _____

SHAYS'S REBELLION

Massachusetts collected taxes on land to pay its war debt. This policy hurt farmers who owned land. The courts made them sell their property to pay taxes. **Daniel Shays** and his followers defied a state order to stop **Shays's Rebellion**. They were defeated by state troops, and 14 leaders were sentenced to death. However, the state freed most, including Shays. Many citizens agreed with Shays.

CALL FOR CHANGE

The weaknesses of the Confederation government led leaders, including James Madison and Alexander Hamilton, to ask all 13 states to send delegates to a Constitutional Convention in Philadelphia in May 1787 to revise the Articles of Confederation, and create a better constitution.

CHALLENGE ACTIVITY

Critical Thinking: Predicting Consider ways in which the new U.S. constitution might change the Articles of Confederation, and list three key changes. **HSS Analysis Skills CS 2, HI 1, HI 2, HI 3**

Chapter 4 Forming a Government

MAIN IDEAS

1. The Constitutional Convention met to improve the government of the United States.

2. The issue of representation led to the Great Compromise.

3. Regional debate over slavery led to the Three-Fifths Compromise.

4. The U.S. Constitution created federalism and a balance of power.

 HSS 8.2

Students analyze the political principles underlying the U.S. Constitution and compare the enumerated and implied powers of the federal government.

Key Terms and People

Constitutional Convention meeting held in Philadelphia to create a new constitution

James Madison a leading convention delegate who was from Virginia

Virginia Plan a plan giving supreme power to the central government and creating a bicameral legislature made of two groups, or houses, of representatives

New Jersey Plan a plan creating a unicameral, or one-house, legislature

Great Compromise it gave each state one vote in the upper house of the legislature and a number of representatives based on its population in the lower house

Three-Fifths Compromise it counted slaves as three-fifths of a person when deciding representation

popular sovereignty the idea that political power belongs to the people

federalism the sharing of power between a central government and the states

legislative branch a Congress of two houses that proposes and passes laws

executive branch the president and the departments that help run the government

judicial branch a system of all the national courts

checks and balances a system that keeps any branch of government from becoming too powerful

Section Summary

CONSTITUTIONAL CONVENTION

The **Constitutional Convention** met in May 1787 in Philadelphia, where America had declared independence. Twelve states sent delegates. Most delegates were educated and had served in state legislatures or Congress. James Madison attended.

> Name one reason Philadelphia was chosen as the site of the Convention.
>
> _____
>
> _____

GREAT COMPROMISE

States disagreed about representation, tariffs, slavery, and strength of the central government. In the **Virginia Plan**, the legislature would be selected on the basis of population. The **New Jersey Plan** proposed that each state receive an equal number of votes. The **Great Compromise** gave every state, regardless of size, an equal vote in the upper house of the legislature. Each state would be represented in the lower house based on population.

> Name the upper house of the federal government created by the Great Compromise.
>
> _____

THREE-FIFTHS COMPROMISE

The **Three-Fifths Compromise** satisfied northerners, who wanted the number of slaves in southern states to determine taxes but not representation. It also satisfied southern delegates, who wanted slaves counted as part of their state populations to increase their power. The delegates agreed to end the slave trade in 20 years.

> Underline the sentence that explains what action about slavery the delegates took.

THE LIVING CONSTITUTION

The delegates wanted to protect **popular sovereignty**. They created **federalism** to accomplish that. States control government functions not assigned to the federal government.

CHECKS AND BALANCES

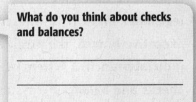

> What do you think about checks and balances?
>
> _____
> _____
> _____

The Constitution balances power among the **legislative branch**, the **executive branch**, and the **judicial branch**. The Constitution's framers established **checks and balances** to prevent any one branch from becoming too strong. The Constitution was completed in September 1787. Congress and then the states ratified the Constitution.

CHALLENGE ACTIVITY

Critical Thinking: Making Judgments Decide whether you support the Three-Fifths Compromise. Give a two-minute speech about your view.
HSS Analysis Skills CS 1, HI 1

Chapter 4 Forming a Government

MAIN IDEAS
1. Federalists and Antifederalists engaged in debate over the new Constitution.
2. The *Federalist Papers* played an important role in the fight for ratification of the Constitution.
3. Ten amendments were added to the Constitution to provide a Bill of Rights to protect citizens.

 HSS 8.2

Students analyze the political principles underlying the U.S. Constitution and compare the enumerated and implied powers of the federal government.

Key Terms and People

Antifederalists people who opposed the Constitution

George Mason delegate who opposed the Constitution

Federalists people who supported the Constitution

Federalist Papers essays supporting the Constitution

amendments official changes to a document

Bill of Rights Constitutional amendments that protect the rights of citizens

Academic Vocabulary

advocate to plead in favor of

Section Summary

FEDERALISTS AND ANTIFEDERALISTS

Antifederalists believed that the Constitutional Convention should not have formed a new government. Delegate **George Mason** opposed the Constitution because it did not contain a section that guaranteed individual rights. Most **Federalists** thought that the Constitution provided a good balance of power. Many wealthy planters, farmers, and lawyers were Federalists. Yet, many craftspeople, merchants, and poor workers also backed the Constitution.

> Why did George Mason oppose the Constitution?
>
> _____
>
> _____
>
> _____

FEDERALIST PAPERS

The *Federalist Papers* were written anonymously by

Section 4, continued

Alexander Hamilton, James Madison, and John Jay
in defense of the Constitution. They tried to per-
suade people that the Constitution would not over-
whelm the states. Madison stated that the diversity
of the United States meant no single group would
take over the government. The Constitution needed
only nine states to pass it, but each state should ratify it as a way of proclaiming national unity. Every
state except Rhode Island held state conventions
that gave citizens the right to discuss and vote on
the Constitution. On December 7, 1787, Delaware
became the first state to ratify it. The Constitution
went into effect in June 1788 after New Hampshire
became the ninth state to ratify it. Several states
ratified the Constitution only after a bill protecting
individual rights was promised.

> Why did states hold constitutional conventions?
> _____
> _____

> What kind of bill did several states demand?
> _____

BILL OF RIGHTS

Many Antifederalists did not believe that the
Constitution would safeguard personal rights.
In the first session of Congress, James Madison
spurred the legislators to develop a bill of rights.
The rights would then become **amendments** to
the Constitution, after a two-thirds majority of
both houses of Congress and three-fourths of the
states approved them. Article V of the Constitution
spelled out this way of changing the document to
bend it to the will of the people.

In December 1791 Congress proposed 12 amendments and turned them over to the states for ratification. By December 1791 the states had ratified
the **Bill of Rights**. Ten of the proposed 12 amendments were written to protect citizens' rights. These
amendments show how to amend the Constitution
to meet the needs of a growing nation.

> Why do you think the Constitution has lasted more than 200 years?
> _____
> _____

CHALLENGE ACTIVITY

Critical Thinking: Comparing and Contrasting Write a short essay comparing and contrasting the views of Federalists and Antifederalists. Use
specific examples. **HSS Analysis Skills HR 5, HI 1**

Chapter 5 Citizenship and the Constitution

HISTORY–SOCIAL SCIENCE STANDARDS
HSS 8.2 Students analyze the political principles underlying the U.S. Constitution and compare the enumerated and implied powers of the federal government.
HSS 8.3 Students understand the foundation of the American political system and the ways in which citizens participate in it.
HSS Analysis Skill HR 5 Students detect the different historical points of view on historical events and determine the context in which the historical statements were made (the questions asked, sources used, author's perspectives).

CHAPTER SUMMARY

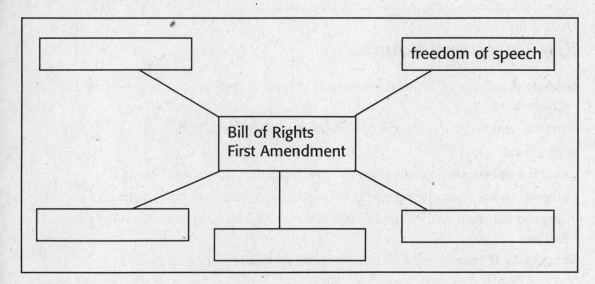

COMPREHENSION AND CRITICAL THINKING

Use the answers to the following questions to fill in the graphic organizer above.

1. Make Inferences How do you think James Madison felt about the possible dangers of majority rule?

_____

_____

2. Elaborate What ways do you think that the First Amendment influences your life?

_____

_____

_____

Chapter 5 Citizenship and the Constitution

MAIN IDEAS

1. The framers of the Constitution devised the federal system.

2. The legislative branch makes the nation's laws.

3. The executive branch enforces the nation's laws.

4. The judicial branch determines whether or not laws are constitutional.

 HSS 8.3

Students analyze the political principles underlying the U.S. Constitution and compare the enumerated and implied powers of the federal government.

Key Terms and People

federal system the government system that gives certain powers to the federal government

impeach vote to bring charges of serious crimes against a president

veto cancel

executive orders commands from the president that have the power of law

pardons orders from the president that grant freedom from punishment

Thurgood Marshall in 1967 he became the first African American Supreme Court Justice

Sandra Day O'Connor the first female Supreme Court Justice, appointed in 1981

Academic Vocabulary

distinct separate

Section Summary

THE FEDERAL SYSTEM

Under the **federal system**, the U.S. Constitution divides powers between the states and the federal government. The Constitution gives the federal government delegated powers, including coining money. It gives state governments or citizens reserved powers, including forming local governments. Concurrent powers are shared by federal and state governments. These powers include taxing. Congress has added powers under the elastic clause to handle new issues.

Describe concurrent powers.

_____

_____

LEGISLATIVE BRANCH

The federal government has three branches. Congress, the legislative branch, has two parts. The House of Representatives has 435 members. The state's population determines the number of representatives for each state. Each state has two senators. They are elected statewide and represent the interests of the entire state.

> **What determines each state's number of representatives?**
> _____
> _____

EXECUTIVE BRANCH

This branch enforces laws made by Congress. The president heads the branch. Americans elect a president every four years. Presidents are limited to two terms. The House of Representatives can **impeach** the president. The Senate tries the cases. Congress dismisses the president if he or she is found guilty. The president and Congress work together. A president can **veto** a law passed by Congress. Congress can undo a veto with a two-thirds majority vote. The president issues **executive orders** to carry out laws affecting the Constitution and other areas. The president also issues **pardons**.

> **How can Congress undo a presidential veto?**
> _____
> _____

JUDICIAL BRANCH

A system of federal courts with the U.S. Supreme Court at the head makes up this branch. Federal courts can undo a state or federal law if the court finds it unconstitutional. Congress can then change the law to make it constitutional. If someone thinks a conviction was unfair, he or she can take the case to the court of appeals. The losing side in that trial may appeal the decision to the U.S. Supreme Court. If the Court declines to hear a case, the court of appeals decision is final. The Supreme Court has become more diverse with the appointments of **Thurgood Marshall** and **Sandra Day O'Connor**.

> **Why might a president appoint more diverse Justices?**
> _____
> _____

CHALLENGE ACTIVITY

Critical Thinking: Judging Would you prefer to serve in the House or the Senate during an impeachment? Write a brief essay explaining why.
HSS Analysis Skills HR 1, HI 2

Chapter 5 Citizenship and the Constitution

MAIN IDEAS

1. The First Amendment guarantees basic freedoms to individuals.
2. Other amendments focus on protecting citizens from certain abuses.
3. The rights of the accused are an important part of the Bill of Rights.
4. The rights of states and citizens are protected by the Bill of Rights.

 HSS 8.3

Students analyze the political principles underlying the U.S Constitution and compare the enumerated and implied powers of the federal government.

Key Terms and People

James Madison Federalist who promised that a Bill of Rights would be added to the Constitution

majority rule the idea that the greatest number of people in a society can make policies for everyone

petition a request made of the government

search warrant an order authorities must get before they search someone's property

due process rule that the law must be fairly applied

indict formally accuse

double jeopardy rule that a person cannot be tried again for the same crime

eminent domain government's power to take personal property to benefit the public

Section Summary

FIRST AMENDMENT

James Madison began writing a list of amendments to the Constitution in 1789. The states ratified 10 amendments, called the Bill of Rights. **Majority rule** could take away smaller groups' rights. The Bill of Rights protects all citizens. First Amendment rights include freedom of religion, freedom of the press, freedom of speech, freedom of assembly, and the right to **petition**. The U.S. government cannot support or interfere with the practice of a religion.

The freedoms of speech and the press give Americans the right to express their own ideas and hear those of others. Freedom of assembly means

What does the Bill of Rights do?
_____
_____

Americans may hold lawful meetings. Citizens can petition for new laws.

PROTECTING CITIZENS

The Second, Third, and Fourth Amendments stem from colonial problems with Britain. The Second Amendment gives state militias the right to bear arms in emergencies. The Third Amendment protects citizens against housing soldiers. The Fourth Amendment protects against certain "searches and seizures." Authorities must obtain a **search warrant** to enter a citizen's property.

> **When is a search warrant needed?**
> _____
> _____

RIGHTS OF THE ACCUSED

The Fifth, Sixth, Seventh, and Eighth Amendments guard the rights of the accused. The Fifth Amendment says that the government cannot take a person's life, liberty, or property without **due process**. A grand jury decides whether to **indict** a person. No one can face **double jeopardy**. Under **eminent domain** the government must pay owners a fair amount for their property. The Sixth Amendment protects an indicted person's rights. The Seventh Amendment says that juries can decide civil cases, usually about money or property. The Eighth Amendment allows bail, or money defendants pay if they fail to appear in court. This amendment also prevents "cruel and unusual punishments" against a person convicted of a crime.

> **Predict a situation in which a government might exercise its right of eminent domain.**
> _____
> _____
> _____

RIGHTS OF STATES AND CITIZENS

The Ninth Amendment states that all citizens' rights are not given by the Constitution. According to the Tenth Amendment, any powers not delegated to the federal government or prohibited to the states are held by the states and the people.

> **What powers are held by the states and the people?**
> _____
> _____

CHALLENGE ACTIVITY

Critical Thinking: Developing In a small group, draw up a new law for which you would like to petition a government official. **HSS Analysis Skills HI 2**

Chapter 5 Citizenship and the Constitution

MAIN IDEAS

1. Citizenship in the United States is determined in several ways.

2. Citizens are expected to fulfill a number of important duties.

3. Active citizen involvement in government and the community is encouraged.

 HSS 8.3
Students understand the foundation of the American political system and the ways in which citizens participate in it.

Key Terms and People

naturalized citizens people who live in the United States whose parents are not citizens and who finally become citizens

deport return to an immigrant's country of origin

draft required military service

political action committees groups that collect money for candidates who support certain issues

interest groups groups of people who share a common interest that motivates them to take political action

Academic Vocabulary

influence change or have an effect on

Section Summary

GAINING U.S. CITIZENSHIP

Naturalized citizens of the United States may become full citizens. First, they apply for citizenship. Then they go through a process that leads to citizenship being granted. At that point, they have most of the rights and responsibilities of other citizens. Legal immigrants have many of those rights and responsibilities but cannot vote or hold public office. The U.S. government can **deport** immigrants who break the law. Legal immigrants over age 18 may seek naturalization after living in the United State five years. After completing the requirements, the person stands before a naturalization court and takes an oath of allegiance to the United States.

> **What limits exist on the rights of legal immigrants?**
> _____
> _____

> **Where is the oath of allegiance taken?**
> _____
> _____

They then receive certificates of naturalization. Two differences between native-born and naturalized citizens are that naturalized citizens can lose their citizenships, and they cannot become president or vice president.

DUTIES OF CITIZENSHIP

Citizens have duties as well as rights. Citizens must obey laws and authority. In addition, they must pay taxes for services, including public roads and public schools. Americans pay a tax on their income to the federal, and sometimes state, government. Men 18 years or older must register with selective service in case of a **draft**. Citizens must serve on juries to give others the right to a trial by jury.

> Why do you think citizens have duties along with rights?
> _____
> _____
> _____

CITIZENS AND GOVERNMENT

Voting in elections is one of a citizen's most important responsibilities. Before voting, a citizen must find out as much as possible about the issues and candidates. A variety of media sources offer information, but some may be deliberately biased. Anyone can help in a campaign, even people not eligible to vote. Many people help with donations to **political action committees** (PACs). People can influence government officials at any time. Many U.S. citizens work with **special interest groups**. Citizens can also work alone in elections or politics in general. Many dedicated Americans also volunteer in community service groups, such as local firefighters or Neighborhood Watch groups that tell police if they see possible criminal activity in their area. Simple acts such as picking up trash in a park or serving food at a food shelter help a community.

> Underline the sentence that mentions media sources.

> Why do you think community service is important and valuable?
> _____
> _____
> _____

CHALLENGE ACTIVITY

Critical Thinking: Analyzing What connects a citizen's rights and responsibilities? Design a graphic organizer showing connections.
HSS Analysis Skills HR 3, HI 1, HI 2

Chapter 6 Launching the Nation

HISTORY–SOCIAL SCIENCE STANDARDS
HSS 8.1 Students understand the major events preceding the founding of the nation and relate their significance to the development of American constitutional democracy.
HSS 8.3 Students understand the foundation of the American political system and the ways in which citizens participate in it.
HSS 8.4 Students analyze the aspirations and ideals of the people of the new nation.
HSS 8.5 Students analyze U.S. foreign policy in the early republic.

CHAPTER SUMMARY

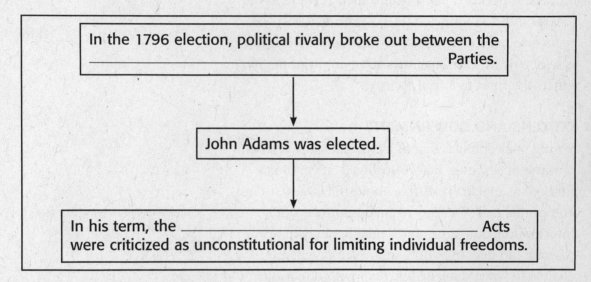

In the 1796 election, political rivalry broke out between the _____ Parties.

John Adams was elected.

In his term, the _____ Acts were criticized as unconstitutional for limiting individual freedoms.

COMPREHENSION AND CRITICAL THINKING

Use the answers to the following questions to fill in the graphic organizer above.

1. Identify Which two parties were rivals in the 1796 presidential election?

_____

2. Describe Which two acts were criticized as unconstitutional during John Adams's presidential term? Describe these acts.

_____

2. Make Judgments If you lived during Adams's term, what would you think about the acts mentioned in Question 2?

_____

_____

Chapter 6 Launching the Nation

MAIN IDEAS

1. In 1789 George Washington became the first president of the United States.

2. Congress and the president organized the executive and judicial branches of government.

3. Americans had high expectations of their new government.

 HSS 8.4
Students analyze the aspirations and ideals of the people of the new nation.

Key Terms and People

George Washington an honest leader, a hero of the revolution, and the first U.S. president

electoral college a group of delegates, or electors, who represent the people's vote in choosing the president

Martha Washington George Washington's wife and the First Lady

precedent an action or a decision that later serves as an example

Judiciary Act of 1789 an act that created three levels of federal courts and defined their powers and relationships to the state courts

Academic Vocabulary

agreement a decision reached by two or more people or groups

Section Summary

THE FIRST PRESIDENT

George Washington was unanimously elected by the **electoral college** in January 1789. John Adams became his vice president. **Martha Washington** entertained and accompanied her husband at social events. She was in charge of the presidential household. Other women, such as Abigail Adams, wife of John Adams, believed women needed to play a larger role in the nation than Martha Washington did. They thought that women should take a more important role in society because they educated their children to be good citizens.

> Why did some women support a larger national role for women?
>
> _____
> _____
> _____

ORGANIZING THE GOVERNMENT

The new federal government had to establish a **precedent** in many areas when creating policies and procedures that would have a great influence on the future of the nation. Congress formed departments in the executive branch to oversee various areas of national policy. Washington consulted with department heads, or cabinet members, who advised him. Our presidents today also meet with their key advisers. Congress passed the **Judiciary Act of 1789**. This act created the federal court system and the courts' location. The president nominated candidates for federal judgeships. The Senate held the power to approve or accept the judges.

> What did cabinet members provide for the president?
> _____

> What limited the president's ability to nominate federal judges?
> _____
> _____
> _____

AMERICANS' EXPECTATIONS OF GOVERNMENT

Americans had high expectations of their government. They wanted trade that did not have the limits put in place by the British Parliament. They also expected the government to protect them and keep the economy on track.

In 1790 4 million people lived in the United States. Most Americans worked on farms. They hoped for fair taxes and the right to move onto western lands. Americans who lived in towns worked as craftspeople, laborers, or merchants. They wanted help with their businesses. Merchants wanted simpler trade laws. Manufacturers wanted laws to shield them from overseas competitors. Most cities were small. Only New York and Philadelphia topped 25,000 residents. New York City, the first capital, reflected the new nation's spirit. In 1792 some 24 Wall Street stockbrokers signed an agreement that eventually created the New York Stock Exchange.

> Underline the sentences that explain the contrast between what country residents and town residents wanted.

> In what year was the agreement signed that led to the New York Stock Exchange?

CHALLENGE ACTIVITY

Critical Thinking: Evaluating You have just attended George Washington's inauguration. Write a letter to a friend describing your thoughts about him. **HSS Analysis Skills CS 1, HR 2, HI 1**

MAIN IDEAS	**HSS 8.3**
1. Hamilton tackled the problem of settling national and state debt. 2. Thomas Jefferson opposed Hamilton's views on government and the economy. 3. Hamilton created a national bank to strengthen the U.S. economy.	Students understand the foundation of the American political system and the ways in which citizens participate in it.

Key Terms and People

Alexander Hamilton the first secretary of the U.S. treasury who wanted to pay the nation's foreign debt immediately and gradually repay the full value of all bonds

national debt money owed by the United States

bonds certificates that represent money

speculators people who buy items at low prices in the hope that the value will rise

Thomas Jefferson the first secretary of state who thought that repaying the full value of all bonds would cheat bondholders who had sold their bonds at low prices

loose construction the view that the federal government can take reasonable actions that the Constitution does not specifically forbid

strict construction the view that the federal government should do only what the Constitution specifically says it can do

The Bank of the United States the national bank

Section Summary

SETTLING THE DEBT

Alexander Hamilton wanted to pay off the **national debt**. He figured that the United States owed $11.7 million to foreign countries. The nation also owed about $40.4 million to U.S. citizens. During the Revolutionary War the government sold **bonds** to raise money. Officials said bonds would be repurchased at a higher price. Some bondholders sold their bonds to **speculators**. Hamilton and **Thomas Jefferson** disagreed on what to do. More politicians agreed with Hamilton. The government replaced old bonds with new, more reliable ones. Hamilton thought that the federal government should repay

> How did the government raise money during the Revolution?
> _____

$21.5 million of the states' debt. But southern lead-
ers objected. Their states had relatively low debts.
Hamilton arranged to have the U.S. capital's location changed from New York to Philadelphia and
finally Washington, D.C. Southern leaders then supported his plan.

> **How did Hamilton persuade the southern leaders?**
> _____
> _____

HAMILTON VERSUS JEFFERSON

Hamilton and Jefferson disagreed about the role of
the central government. Hamilton wanted a strong
federal government. Jefferson wanted strong powers for the states. Hamilton did not want people to
have much power because he had little faith in the
average person. Jefferson believed that the people
had the right to rule the country. Hamilton backed
manufacturing and business, and higher tariffs.
Jefferson backed farming, and lower tariffs.

> **Underline the sentences that explain Hamilton's and Jefferson's views of the American people.**

THE DEBATE OVER THE BANK

In 1791 Hamilton and Jefferson disagreed about the
government's economic problems. Hamilton wanted a national bank, so the government could safely
deposit money. Jefferson believed that Hamilton's
plan gave too much power to the federal government. Hamilton supported **loose construction** of
the Constitution. Jefferson backed **strict construction**. Washington and Congress wanted the **Bank of
the United States**. It helped make the U.S. economy
more stable.

> **Name one reason Hamilton supported a national bank.**
> _____
> _____

> **Do you think the Bank was a good idea? Why or why not?**
> _____
> _____

CHALLENGE ACTIVITY

Critical Thinking: Evaluating Do you think Hamilton or Jefferson was
more correct in his views of people? Give a brief speech explaining your
opinion. **HSS Analysis Skills HR 1, HR 2, HI 1, HI 6**

Chapter 6 Launching the Nation

Section 3

MAIN IDEAS

1. The United States tried to remain neutral regarding events in Europe.
2. Washington's government settled conflicts at home.
3. In his Farewell Address, President Washington gave advice to the nation.

 HSS 8.5
Students analyze U.S. foreign policy in the early republic.

Key Terms and People

French Revolution a rebellion of the French people against their king that led to the creation of a republican government

Neutrality Proclamation stated that the United States would not take sides with any European countries who were at war

privateers private ships hired by a country to attack its enemies

Jay's Treaty settled the disputes that had arisen between the United States and Britain in the early 1790s

Pinckney's Treaty settled border and trade disputes with Spain

Little Turtle an American Indian chief who led forces to defeat U.S. forces in 1790

Battle of Fallen Timbers the battle that broke the strength of Indian forces in the Northwest Territory

Treaty of Greenville gave the United States right of entry to American Indian lands

Whiskey Rebellion an uprising in which some farmers refused to pay the whiskey tax

Academic Vocabulary

neutral unbiased, not favoring either side in a conflict

Section Summary

REMAINING NEUTRAL

The **French Revolution** increased tensions between France and Britain. Many Americans supported the French Revolution, but others opposed it. France and Great Britain finally went to war. George Washington stated U.S. neutrality toward the war in the **Neutrality Proclamation**. A French representative asked American sailors to command **privateers** to aid France in fighting England. Washington said that this violated

> Why do you think some Americans supported the French Revolution?
>
> _____
>
> _____

U.S. neutrality. Jefferson thought the United States should support France and resented interference in his role as secretary of state. He resigned in 1793.

Washington wanted to stop a war between the United States and Britain. The two sides signed **Jay's Treaty**. Britain would pay damages on seized American ships. Spain and the United States disputed the border of Florida. **Pinckney's Treaty** settled that issue and reopened New Orleans to American ships.

> What did the United States gain from Pinckney's Treaty?
> _____
> _____

CONFLICT IN THE NORTHWEST TERRITORY

Americans continued to settle the territory despite protests of American Indians. U.S. forces lost a battle to Miami chief **Little Turtle**. But General Anthony Wayne commanded U.S. troops in gaining the territory at last. The American Indians were defeated in the **Battle of Fallen Timber**.

> Why might Americans Indians have protested the U.S. settlements?
> _____
> _____

THE WHISKEY REBELLION

In March 1791 Congress passed a tax on American-made whiskey. The **Whiskey Rebellion** broke out. Washington personally led the army against the rebels in western Pennsylvania, but they fled. The revolt ended with no battle.

WASHINGTON SAYS FAREWELL

Washington declined to run for a third term. He had tired of public life and considered the American people the nation's leaders. In his farewell speech, he warned about the dangers of foreign ties and political conflicts at home. He also cautioned against too much debt. At the conclusion of his speech, he stated that he looked forward to a life "of good laws under a free government. . ."

> Name two dangers that Washington mentioned.
> _____
> _____

CHALLENGE ACTIVITY

Critical Thinking: Sequencing Create a time line of important events in the 1790s. Describe each of the events and, when appropriate, explain how one event caused or resulted from another event. Illustrate your time line. **HSS Analysis Skills CS 2, HI 2**

Chapter 6 Launching the Nation

MAIN IDEAS

1. The rise of political parties created competition in the election of 1796.

2. The XYZ affair caused problems for President John Adams.

3 Controversy broke out over the Alien and Sedition Acts.

 HSS 8.5
Students understand the foundations of the American political system and the ways in which citizens participate in it.

Key Terms and People

political parties groups that help elect people and shape politics

Federalist Party wanted a strong federal government and supported industry and trade

Democratic-Republican Party wanted to limit the federal government's powers

XYZ affair a French demand for a $250,000 bribe and a $12 million loan in exchange for a treaty

Alien and Sedition Acts acts that punished supporters of France and deprived people of the freedom to say and write what they believed.

Kentucky and Virginia Resolutions said that the Alien and Sedition Acts were unconstitutional

Section Summary

THE ELECTION OF 1796

In the election of 1796, more than one candidate ran for president. **Political parties** had started during Washington's presidency. Washington cautioned against party rivalry in his farewell, but rivalry dominated the 1796 election. Alexander Hamilton was key in founding the **Federalist Party**. John Adams and Thomas Pinckney were the Federalist candidates. Thomas Jefferson and James Madison helped found the **Democratic-Republican Party**. That party selected Thomas Jefferson and Aaron Burr as its candidates. Business people in cities tended to support Adams. Farmers generally favored Jefferson. Both sides attacked each other. Adams won; Jefferson was second. He and Jefferson then had to serve as president and vice president.

> **Who helped start the Federalist Party?**
> _____
> _____

PRESIDENT ADAMS AND THE XYZ AFFAIR

Adams made improving the relationship between the United States and France a high priority. France was unhappy when the United States refused to let its citizens join in the war against Britain. Adams sent U.S. diplomats to repair that problem and make a treaty to guard U.S. shipping. The French foreign minister refused to meet with them. Three French agents said that the minister would discuss a treaty only if America paid a $250,000 bribe and gave a $12 million loan. The American public became furious about the **XYZ affair**. Still, Adams did not declare war on France. This angered many other Federalists. At last the United States and France did negotiate a peace treaty.

> In the end, what occurred between the United States and France?
> _____
> _____

THE ALIEN AND SEDITION ACT

The **Alien and Sedition Acts**, passed by Federalists in Congress, became law in 1798. The Alien Act empowered the president to remove foreign residents he decided were involved in any treasonable or secret plots against the government. The Sedition Act forbid U.S. residents to "write, print, utter, or publish" false or critical words against the government. The **Kentucky and Virginia Resolutions** stated that the acts were unconstitutional. Jefferson and James Madison said that the states could disobey unconstitutional federal laws. Congress did not repeal the acts, though they were not renewed. The resolutions presented the view that states could dispute the federal government. Later politicians would say this idea meant that the states could declare laws or actions of the federal government to be illegal.

> Underline the sentence that explains what the Sedition Act did.

CHALLENGE ACTIVITY

Critical Thinking: Predicting Do some research to discover when in U.S. history states would say that they could declare federal law to be illegal. Write a brief essay explaining both sides of the argument.

HSS Analysis Skills CS 1, CS 5, HI 2, HI 3

Chapter 7 The Jefferson Era

HISTORY–SOCIAL SCIENCE STANDARDS
HSS 8.4 Students analyze the aspirations and ideals of the people of the new nation.
HSS 8.5 Students analyze U.S. foreign policy in the early Republic.
HSS 8.8 Students analyze the divergent paths of the American people in the West from 1800 to the mid-1800s and the challenges they faced.

CHAPTER SUMMARY

Analyzing the War of 1812	
Reasons U.S. Entered War	1. British trade restrictions 2. 3.
Reasons U.S. Might Win	1. British at war with the French at the same time 2.
Reasons British Might Win	1. 2. White House burned

COMPREHENSION AND CRITICAL THINKING

Use the answers to the following questions to fill in the graphic organizer above.

1. Explain Why did the United States go to war with Britain in 1812?

_____

_____

2. Evaluate If you were living at the time the war started, would you have felt confident that the United States would win the war? Why or why not?

_____

_____

3. Draw a Conclusion How did the War of 1812 help the United States?

_____

_____

Chapter 7 The Jefferson Era

MAIN IDEAS

1. The election of 1800 marked the first peaceful transition in power from one political party to another.

2. President Jefferson's beliefs about the federal government were reflected in his policies.

3. *Marbury* v. *Madison* increased the power of the judicial branch of government.

 HSS 8.4
Students analyze the aspirations and ideals of the people of the new nation.

Key Terms and People

John Adams Federalist president first elected in 1796 who lost the 1800 presidential election

Thomas Jefferson Republican who defeated John Adams in the presidential election of 1800

John Marshall a Federalist appointed by Adams to be Chief Justice of the Supreme Court

Marbury* v. *Madison a case that established the Supreme Court's power of judicial review

judicial review the Supreme Court's power to declare an act of Congress unconstitutional

Academic Vocabulary

functions uses or purposes

Section Summary

THE ELECTION OF 1800

Thomas Jefferson defeated **John Adams** and became president in 1800. In campaigning, both sides had made their cases in newspaper editorials and letters. Both sides believed that if the other gained power, the nation would be destroyed. The campaigning was intense. Federalists said if Jefferson gained power, revolution and chaos would follow. Republicans claimed that Adams would crown himself king. Jefferson and Aaron Burr, his vice presidential running mate, each won 73 votes. After the thirty-sixth ballot in the House of Representatives, Jefferson was elected President.

How did the presidential candidates wage the campaign of 1800?

_____

_____

JEFFERSON'S POLICIES

Jefferson gave his first speech in the new capitol.
He said he supported the will of the majority. He
emphasized his belief in a limited government and
the protection of civil liberties. Jefferson convinced
Congress to let the Alien and Sedition Acts expire.
He cut military spending to free money to pay the
national debt. The Republican-led Congress passed
laws to end the unpopular whiskey tax and other
domestic taxes.

> **Name one action Jefferson took based on his principles.**
> _____
> _____

In 1801 the national government was made up
of only several hundred people. Jefferson liked it
that way. He thought that safeguarding the nation
against foreign threats, delivering the mail, and
collecting custom duties were the most important
functions of the federal government. Jefferson had
fought Alexander Hamilton over the creation of the
Bank of the United States, but he did not close it.

> **Why might Jefferson have left the national bank alone?**
> _____
> _____

MARBURY V. MADISON

Adams filled 16 new federal judgeships with
Federalists before leaving office. Republicans in
Congress soon repealed the Judiciary Act upon
which Adams's appointments were based. A controversy arose when Adams appointed William
Marbury as a justice of the peace. The documents
supporting Marbury's appointment were never
delivered. When Jefferson took office, Secretary
of State James Madison would not deliver them.
Marbury sued and asked the Supreme Court to
order Madison to give him the documents. **John
Marshall** wrote the Court's opinion in *Marbury* v.
Madison. He ruled that the law which Marbury's
case depended upon was unconstitutional. The case
established the Court's power of **judicial review**.

> **Why did Marbury sue Madison?**
> _____
> _____

CHALLENGE ACTIVITY

Critical Thinking: Making Inferences What if the 1800 campaign were
waged as campaigns are waged now. Write a speech that you think Thomas
Jefferson would give. Deliver his speech. **HSS Analysis Skills HI 5**

Chapter 7 The Jefferson Era

MAIN IDEAS

1. As American settlers moved West, control of the Mississippi River became more important to the United States.

2. The Louisiana Purchase almost doubled the size of the United States.

3. Expeditions led by Lewis, Clark, and Pike increased Americans' understanding of the West.

 HSS 8.8

Students analyze the divergent paths of the American people in the West from 1800 to the mid-1800s and the challenges they faced.

Key Terms and People

Louisiana Purchase the U.S. government paid $15 million to France for Louisiana, which roughly doubled the size of the United States

Meriwether Lewis a former army captain Jefferson chose to lead an expedition to explore the West

William Clark co-leader of the western expedition

Lewis and Clark expedition a long journey to explore the Louisiana Purchase

Sacagawea a Shoshone Indian who helped the expedition by naming plants and gathering edible fruits and vegetables for the group

Zebulon Pike he explored the West to find the Red River and reached the summit of the mountain now known as Pike's Peak

Section Summary

AMERICAN SETTLERS MOVE WEST

Thousands of Americans moved into the area between the Appalachians and the Mississippi River. The settlers used the Mississippi and Ohio rivers to move their products to eastern markets. Jefferson was concerned that a foreign power might shut down the port of New Orleans, which settlers needed to move their goods east and to Europe. Spain governed New Orleans and Louisiana, which extended from the Mississippi to the Rocky Mountains. Under a secret treaty, Spain gave Louisiana to France, transferring the problem of trying to keep Americans out of Louisiana.

> **Why did Jefferson worry about the port of New Orleans?**
> _____
> _____

> **Why would Americans want to move into Louisiana?**
> _____
> _____

LOUISIANA PURCHASE

In 1802, before giving Louisiana to France, Spain shut American shipping out of New Orleans. Jefferson sent U.S. representatives to France to buy New Orleans. Napoleon ruled France. He wanted to rebuild France's empire in North America. But Napoleon had no base for a conquest of Louisiana. He also needed money to wage war against Great Britain. The United States bought the western territory for $15 million in the **Louisiana Purchase**.

> **What are two reasons that Napoleon did not try to conquer Louisiana?**
> _____
> _____

EXPEDITIONS UNCOVER THE WEST

Western Native Americans and the land they lived on were a mystery to others. President Jefferson wanted to know about them and their land. He also wondered if there was a river route to the Pacific Ocean. In 1803 Congress provided money to explore the West. **Meriwether Lewis** and **William Clark** were chosen to lead the **Lewis and Clark expedition**, which began in May 1804. Lewis and Clark and their crew traveled up the Missouri River. Finally, they saw Native Americans, and Lewis used interpreters to tell their leaders that the United States now owned the land on which they lived. **Sacagawea** and her husband aided Lewis and Clark. Lewis and Clark did not find a river route to the Pacific, but they learned much about western lands.

> **Underline the sentences that explain why Jefferson wanted to know more about the West.**

In 1806 **Zebulon Pike** was sent to locate the Red River, which was the Louisiana Territory's border with New Spain. In present-day Colorado he reached the summit of Pike's Peak. Spanish cavalry arrested him in Spanish-held lands and imprisoned him. When released he returned to the United States and reported on his trip. He gave many Americans their first information about the Southwest.

> **What are some of the difficulties faced by American expeditions in the West?**
> _____
> _____

CHALLENGE ACTIVITY

Critical Thinking: Drawing Inferences Some members of the Lewis and Clark expedition kept journals or diaries. Write a brief diary entry as if you were a member of the expedition. **HSS Analysis Skills CS 2**

MAIN IDEAS
1. Violations of U.S. neutrality led Congress to enact a ban on trade.
2. Native Americans, Great Britain, and the United States came into conflict in the West.
3. The War Hawks led a growing call for war with Great Britain.

 HSS 8.5
Students analyze U.S. foreign policy in the early Republic.

Key Terms and People

USS *Constitution* a large U.S. warship sent to end attacks by Mediterranean pirates on American merchant ships

impressment practice of forcing people to serve in the army or navy

embargo the banning of trade

Embargo Act a U.S. law that essentially banned trade with all foreign countries

Non-Intercourse Act a new law banning trade only with Great Britain, France, and their colonies

Tecumseh a brilliant speaker who warned other Native Americans that settlers wanted their lands

Battle of Tippecanoe the battle between the U.S. forces and Tecumseh's followers that ended with the U.S. forces winning

War Hawks several members of Congress who called for war against Great Britain

James Madison a Republican who was elected president in 1808

Section Summary

VIOLATIONS OF NEUTRALITY

In the late 1700s and early 1800s, American merchant ships sailed the oceans. The profitable overseas trade was dangerous. Pirates seized cargo and held crews for ransom. The United States sent the **USS *Constitution*** and other ships to end the attacks. When Great Britain and France declared war in 1803, each tried to stop the United States from selling goods to the other. The British and French searched many American ships for war goods. Then Britain started searching American ships for sailors who had deserted

> **Why did Britain and France try to stop the United States from selling goods to the other?**
> _____
> _____

the British navy. At times U.S. citizens were seized by accident.

Impressment continued over U.S. protests. Thomas Jefferson, who had been re-elected in 1804, favored an **embargo** rather than war with Britain. In late 1807, Congress passed the **Embargo Act** to punish Britain and France. Merchants lost huge amounts of money because of the act. In 1809, Congress replaced the embargo with the **Non-Intercourse Act**. That law did not work either.

> **How was an embargo an alternative to war?**
> _____
> _____

CONFLICT IN THE WEST

In the West, Native Americans, the United States, and Great Britain clashed. As settlers poured into the West, Native Americans lost land that they believed was unfairly taken. British agents from Canada armed Native Americans in the West. **Tecumseh**, a Shawnee chief, united his forces with the Creek nation. William Henry Harrison, the governor of the Indiana Territory, raised an army to battle him. At the day-long **Battle of Tippecanoe**, Harrison's forces defeated the Native Americans.

> **How did British agents aid Native Americans in the West?**
> _____
> _____

CALL FOR WAR

War Hawks in Congress led in demanding war against Britain. The leaders wanted to end British influence on Native Americans. They resented British restraints on U.S. trade. Others opposed war against Britain. They believed America lacked the military strength to win. In 1808 Republican **James Madison** was elected president. He had difficulty carrying on the unpopular trade policy. In 1812 he asked Congress to vote on whether to wage war against Britain. Congress voted to declare war. Madison was again elected. He became commander in chief in the War of 1812.

> **Describe the problem that Madison faced in 1808.**
> _____
> _____
> _____

CHALLENGE ACTIVITY

Critical Thinking: Predicting What difficulties might the United States face in the War of 1812? List them. **HSS Analysis Skills HI 3**

Section 4

MAIN IDEAS

1. American forces held their own against the British in the early battles of the war.
2. The effects of the war included prosperity and national pride.

 HSS 8.5

Students analyze U.S. foreign policy in the early Republic.

Key Terms and People

Oliver Hazard Perry U.S. Navy commodore who won a victory against the British

Battle of Lake Erie the victory won by Perry and his sailors

Andrew Jackson the commander of the Tennessee militia who led an attack on the Creek nation in Alabama

Treaty of Fort Jackson the treaty that forced the Creek nation to give up millions of acres of their land

Battle of New Orleans the last major conflict of the War of 1812, which made Andrew Jackson a hero

Hartford Convention a meeting of Federalists opposed to the war

Treaty of Ghent the pact that ended the War of 1812

Academic Vocabulary

consequences the effects of a particular event or events

Section Summary

EARLY BATTLES

In 1812 the United States launched a war against a dominant nation. The British navy had hundreds of ships. The U.S. Navy had fewer than 20 ships, but it boasted expert sailors and big new warships. American morale rose when its ships defeated the British in several battles. Finally, the British blockaded U.S. seaports.

> Underline the sentences that contrast the U.S. and British navies.

The U.S. planned to attack Canada from Detroit, from Niagara Falls, and from the Hudson River Valley toward Montreal. British soldiers and Native Americans led by Tecumseh took Fort Detroit. State militia doomed the other two attacks against Canada by arguing that they were not required to fight in a foreign country.

In 1813 the United States planned to end Britain's rule of Lake Erie. Commodore **Oliver Hazard Perry** and his small fleet won the **Battle of Lake Erie**. General Harrison then marched his troops into Canada. He defeated a combined force of British and Native Americans, breaking Britain's power. Tecumseh died in the fighting, harming the alliance of the British and the Native Americans.

> What effect did the death of Tecumseh have?
> _____
> _____

In 1814 **Andrew Jackson** won a battle against the Creek nation that ended in the **Treaty of Fort Jackson**.

GREAT BRITAIN ON THE OFFENSIVE

The British sent more troops to America after defeating the French in 1814. The British set fire to the White House and other buildings in Washington, D.C. The British also attacked New Orleans. Andrew Jackson commanded forces made up of regular soldiers. His forces included two battalions of free African Americans, a group of Choctaw Indian militia, and pirates led by Jean Lafitte. Although Jackson's forces were outnumbered, America won the **Battle of New Orleans**, the last key battle of the war. Andrew Jackson became a war hero.

> Why was Andrew Jackson considered a hero?
> _____
> _____

EFFECTS OF THE WAR

Before Federalist delegates from the **Hartford Convention** reached Washington, the war had ended. Slow communications meant that neither Jackson nor the Federalists heard that the **Treaty of Ghent** finished the war. Each nation gave back the territory it had conquered. Yet the war had led to intense patriotism in America as well as growth in American manufacturing.

> In what ways did the war benefit the United States?
> _____
> _____
> _____

CHALLENGE ACTIVITY

Critical Thinking: Drawing Inferences You are the first mate on a New England trading ship several months after the War of 1812. Write a letter about how the end of the war affects you. **HSS Analysis Skills HI 1**

Chapter 8 A New National Identity

HISTORY–SOCIAL SCIENCE STANDARDS
HSS 8.4 Students analyze the aspirations and ideals of the people of the new nation.
HSS 8.5 Students analyze U.S. Foreign policy in the early Republic.
HSS 8.6 Students analyze the divergent paths of the American people from 1800 to the mid-1800s and the challenges they faced, with emphasis on the Northeast.

CHAPTER SUMMARY

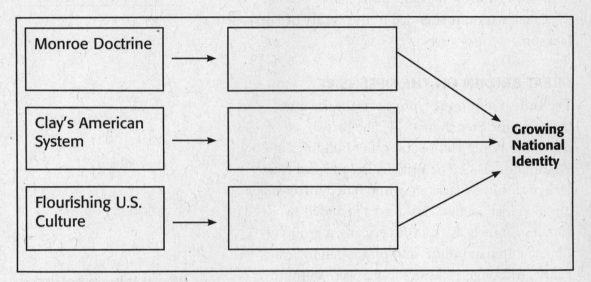

COMPREHENSION AND CRITICAL THINKING

Use the answers to the following questions to fill in the graphic organizer above.

1. Identify Cause and Effect How did the Monroe Doctrine contribute to America's growing national identity?

_____

_____

2. Identify Cause and Effect What was the goal of Henry Clay's American System?

_____

_____

3. Draw Conclusions In what ways do you think the growing contributions of U.S. culture increased the sense of national identity?

_____

_____

Chapter 8 A New National Identity

MAIN IDEAS

1. The United States and Great Britain settled their disputes over boundaries and control of waterways.

2. The United States gained Florida in an agreement with Spain.

3. With the Monroe Doctrine, the United States strengthened its relationship with Latin America.

 HSS 8.5
Students analyze U.S. foreign policy in the early Republic.

Key Terms and People

Rush-Bagot Agreement a compromise that limited U.S. and British naval power on the Great Lakes

Convention of 1818 a treaty that gave the United States fishing rights off parts of the Newfoundland and Labrador coasts

James Monroe U.S. president elected in 1816

Adams-Onís Treaty an agreement that settled all border disputes between the United States and Spain

Simon Bolívar the leader of the successful revolutions of Latin American colonies against Spain

Monroe Doctrine an exclusive statement of American policy warning European nations not to interfere with the Americas

Academic Vocabulary

circumstances surrounding situation

Section Summary

SETTLING DISPUTES WITH GREAT BRITAIN

After the War of 1812 ended, both the United States and Great Britain wanted to retain their navies and freedom to fish on the Great Lakes. The **Rush-Bagot Agreement** resolved that issue. The **Convention of 1818** gave America certain fishing rights, and it established the border between the United States and Canada. In this treaty, both countries agreed to occupy the Pacific Northwest together.

> What were the results of the Convention of 1818?
>
> _____
>
> _____

THE UNITED STATES GAINS FLORIDA

The United States also debated its border with
Spanish Florida. President **James Monroe** sent
General Andrew Jackson and troops to protect the
U.S.-Florida border. Seminole Indians often aided
runaway slaves and sometimes raided U.S. settlements. Under Jackson's command, U.S. troops
invaded Florida to catch Seminole raiders, starting
the First Seminole War.

> **Why did President Monroe send Jackson to Florida?**
> _____
> _____

Jackson also captured most of Spain's key military posts. Jackson took these actions without a
direct command from the president. The Spanish
were upset, but most Americans backed Jackson.
In 1819, Secretary of State John Quincy Adams
and Spanish diplomat Luis de Onís negotiated the
Adams-Onís Treaty. This treaty settled all border
disputes between the United States and Spain.

> **What convinced the Spanish to negotiate with the Americans?**
> _____
> _____

THE MONROE DOCTRINE

By the early 1820s most Latin American countries
had won independence from Spain. **Simon Bolívar**,
called the Liberator, led many of these battles. The
United States saw the struggles as comparable to the
American Revolution. United States leaders supported the Latin Americans in their struggles with
European powers.

> **Why do you think the United States supported Latin American independence?**
> _____
> _____

Monroe developed the **Monroe Doctrine** to
guard against European countries interfering
with the new Latin American nations. The document spells out the relationship between European
nations and the United States in the Western
Hemisphere. The doctrine states that the United
States will intervene in Latin American affairs when
its own security is at risk. Few European nations
challenged the doctrine.

CHALLENGE ACTIVITY

Critical Thinking: Cause and Effect Make a chart identifying the causes
and effects of the Rush-Bagot Agreement, the Adams-Onís Treaty, and
the Monroe Doctrine. **HSS Analysis Skill HI 2**

Chapter 8 A New National Identity

MAIN IDEAS

1. Growing nationalism led to improvement in the nation's transportation systems.
2. The Missouri Compromise settled an important regional conflict.
3. The outcome of the election of 1824 led to controversy.

 HSS 8.4
Students analyze the aspirations and ideals of the people of the new nation.

Key Terms and People

nationalism a sense of pride and devotion to a nation

Henry Clay a U.S. representative from Kentucky who supported an emphasis on national unity

American System a series of measures intended to make the United States economically self-sufficient

Cumberland Road the first road built by the federal government

Erie Canal a waterway that ran from Albany to Buffalo, New York

Era of Good Feelings a U.S. era of peace, pride, and progress

sectionalism disagreement between leaders of different regions

Missouri Compromise an agreement that settled the conflict over Missouri's application for statehood

John Quincy Adams chosen as president by the House of Representatives in 1824

Academic Vocabulary

incentive something that leads people to follow a certain course of action

Section Summary

GROWING NATIONALISM

Americans appreciated a rising sense of **nationalism** based on favorable negotiations with foreign nations. **Henry Clay** firmly supported this nationalism. Clay developed the **American System** to help create a stronger national economy and reduce regional disagreements. He pushed for a protective tariff and a national bank that would back a single currency to encourage interstate trade. The tariff funds could help improve roads and canals.

> How would a single currency encourage interstate trade?
> _____
> _____

The mainly dirt roads in the United States made travel hard in the early 1800s. The **Cumberland Road** stretched from Cumberland, Maryland, to Wheeling, on the Ohio River in present-day West Virginia. By 1850 its extension reached Illinois.

Building of the **Erie Canal** started in 1817 and was finished in 1825. Water transportation was often faster, less expensive, and easier than road travel. British, German, and Irish immigrants dug the entire canal by hand. At that time, the United States enjoyed the **Era of Good Feelings**.

> Why do you think canal transportation was an important innovation?
> _____
> _____

THE MISSOURI COMPROMISE

Disagreements between the North and South, known as **sectionalism**, threatened the Union. When Missouri applied to enter the Union, the Union contained 11 free states and 11 slave states. The Senate's balance would favor the South if Missouri entered as a slave state. Henry Clay persuaded Congress to agree to the **Missouri Compromise**. Missouri entered the Union as a slave state, and Maine entered as a free state. This kept an equal balance in the Senate. Slavery was banned in new territories or states north of Missouri's southern border.

> How did Henry Clay help Missouri enter the Union?
> _____
> _____

THE ELECTION OF 1824

Senator Andrew Jackson gained the most popular votes, but not enough electoral votes to win the election. The House of Representatives chose **John Quincy Adams** as president. Jackson's supporters claimed that Adams had made a "corrupt bargain" with Representative Henry Clay to win. Later, Adams named Clay secretary of state. The election controversy cost Adams support among Americans.

> How did Andrew Jackson lose the election of 1824?
> _____
> _____

CHALLENGE ACTIVITY

Critical Thinking: Drawing Conclusions Would you support Adams despite the Clay controversy? Why or why not? **HSS Analysis Skill HI 1**

Chapter 8 A New National Identity

MAIN IDEAS

1. American writers created a new style of literature.
2. A new style of art showcased the beauty of America and its people.
3. American ideals influenced other aspects of culture, including religion and music.
4. Architecture and education were affected by cultural ideals.

 HSS 8.4
Students analyze the aspirations and ideals of the people of the new nation.

Key Terms and People

Washington Irving one of the first American writers to gain international fame

James Fenimore Cooper perhaps the best known of the new American writers

Hudson River school a group of artists whose paintings reflected national pride and an appreciation of the American landscape

Thomas Cole a landscape painter who was a founder of the Hudson River school

George Caleb Bingham an artist whose paintings showed both the American landscape and scenes from people's daily lives

Section Summary

AMERICAN WRITERS

Americans expressed their thoughts and feelings in literature and art. They took spiritual comfort in religion and music. The strengthening national identity was shown in education and architecture.

Washington Irving often wrote about American history. He cautioned Americans to learn from the past and prepare for the future. He often used a humorous style of writing called satire. In "Rip Van Winkle," one of his most famous short stories, Irving expresses his idea that Americans must use past lessons to deal with the future.

James Fenimore Cooper wrote about characters who lived on the frontier, including Native Americans. By placing some characters in historical events, he popularized historical fiction.

What was Irving's message?
_____
_____

Interactive Reader and Study Guide

A NEW STYLE OF ART

The works of Irving and Cooper inspired painters. By the 1830s the **Hudson River school** had
appeared. **Thomas Cole** portrayed the American
landscape's unique traits. Other painters followed
his lead. **George Caleb Bingham** created a painting
that shows the rough lives of western traders as well
as the landscape.

> Why did Bingham show the West
> as a rough place to live?
> _____
> _____

RELIGION AND MUSIC

Religious revivalism fanned out across America
through the early and mid-1800s. Leaders met with
large crowds to reawaken religious faith. People
sang songs known as spirituals at revival meetings.
Spirituals are a kind of folk hymn from both white
and African American music traditions. Popular
folk music showed the unique views of the nation.
"Hunters of Kentucky" honored the Battle of New
Orleans. It was used successfully in the presidential
campaign of Andrew Jackson in 1828.

> How were spirituals important for
> religious revival meetings?
> _____
> _____

ARCHITECTURE AND EDUCATION

In pre-Revolution America, most American architects modeled their designs on the style used
in Great Britain. After the Revolution, Thomas
Jefferson said that Americans should base their
building designs on those of ancient Greece and
Rome. Many architects agreed with Jefferson and
used Greek and Roman styles.

> Why do you think Jefferson
> wanted Americans to change their
> styles of architecture?
> _____
> _____

Americans also found education important. In
1837 Massachusetts set up a state board of education. Other states followed Massachusetts and started their own education systems.

CHALLENGE ACTIVITY

Critical Thinking: Elaborating Would you prefer to be a writer or a
landscape artist? Make a choice, and either write an essay describing the
western American landscape in the 1800s or draw a picture of it.
HSS Analysis Skill CS 3

Chapter 9 The Age of Jackson

HISTORY–SOCIAL SCIENCE STANDARDS

HSS 8.4 Students analyze the aspirations and ideals of the people of the new nation.

HSS 8.8 Students analyze the divergent paths of the American people in the West from 1800 to the mid-1800s and the challenges they faced.

HSS 8.10 Students analyze the multiple causes, key events, and complex consequences of the Civil War.

CHAPTER SUMMARY

CAUSE		EFFECT
More people received the right to vote	⟶	
	⟶	Northerners supported tariffs
	⟶	Southerners opposed tariffs
The United States wanted to control more land	⟶	Indian Removal Act passed in 1830

COMPREHENSION AND CRITICAL THINKING

Use the answers to the following questions to fill in the graphic organizer above.

1. Explain Give one reason that the political system changed in the early 1800s.

_____

_____

2. Identify Cause and Effect Why did most northerners support tariffs and most southerners oppose them?

_____

_____

Chapter 9 The Age of Jackson

MAIN IDEAS

1. The 1820s saw an expansion of democracy as more Americans held the right to vote.

2. Jackson's victory in the election of 1828 marked a change in American politics.

 HSS 8.4
Students analyze the aspirations and ideals of the people of the new nation.

Key Terms and People

nominating conventions public meetings to select a party's presidential and vice presidential candidates

Jacksonian Democracy the democratic expansion that occurred during Jackson's presidency

Democratic Party a party formed by Jackson supporters

John C. Calhoun Jackson's vice presidential running mate

spoils system the practice of rewarding political supporters with government jobs

Martin Van Buren the secretary of state in Jackson's cabinet

Kitchen Cabinet an informal group of Jackson's trusted advisers that sometimes met in the White House kitchen

Section Summary

A CHANGING ELECTORATE

In the early 1800s, state legislatures expanded democracy, giving more people voting rights. However, women and African Americans still had no voting rights in most states. By 1828 almost all states had changed the system under which state legislatures nominated electors in the electoral college. Now, the people nominated their own electors. Some parties began to hold **nominating conventions**. Broader voting rights and conventions allowed more people to actively participate in politics.

> Who was left out in the push to give Americans more voting rights?
>
> _____
>
> _____

JACKSONIAN DEMOCRACY

Andrew Jackson entered the political scene as American democracy grew. Historians called the expansion of democracy in this era **Jacksonian Democracy.** Jackson's supporters were mainly farmers, frontier settlers, and southern slaveholders. They believed he would protect the rights of the common people and the slave states. They referred to themselves as Democrats and established the **Democratic Party.** Many supporters of President John Quincy Adams called themselves National Republicans.

> **Who do you think supported Adams for president?**
> _____
> _____

THE 1828 ELECTION

The presidential candidates were President Adams and Andrew Jackson in a replay of the 1824 election. Jackson selected South Carolina Senator **John C. Calhoun** as his running mate. The campaign concentrated on personalities. Jackson's campaigners said he was a war hero who was born poor and earned success through hard work. They said that Adams knew nothing about everyday people because his father had been the second U.S. president. Adams's backers said Jackson was too coarse to be president.

> **What were some of the key differences between Jackson and Adams?**
> _____
> _____

Jackson and Calhoun won the election. Jackson's supporters described his victory as a triumph for the common people. A crowd of some 20,000 people held a big party on the White House lawn to celebrate. Jackson began the **spoils system**, but he replaced fewer than one-fifth of federal officeholders. One of Jackson's strongest cabinet members was **Martin Van Buren.** Jackson also relied heavily on a trusted group of advisors that was called the **kitchen cabinet.**

> **Why might so many people have attended the election party?**
> _____
> _____
> _____

CHALLENGE ACTIVITY

Critical Thinking: Analyze Make a chart contrasting people's opinions about Adams and Jackson. **HSS Analysis Skill CR 2.**

Chapter 9 The Age of Jackson

MAIN IDEAS

1. Regional differences grew during Jackson's presidency.
2. The rights of the states were debated amid arguments about a national tariff.
3. Jackson's attack on the Bank sparked controversy.
4. Jackson's policies led to the Panic of 1837.

 HSS 8.10
Students analyze the multiple causes, key events, and complex consequences of the Civil War.

Key Terms and People

Tariff of Abominations a tariff with very high rates

states' rights doctrine the belief that state power should be greater than federal power

nullification crisis the dispute over whether states had the right to nullify, or disobey, any federal law with which they disagreed

Daniel Webster a senator from Massachusetts who spoke out against nullification and believed the nation had to stay united

McCulloch v. *Maryland* the case in which the U.S. Supreme Court ruled that the Second Bank of the United States was constitutional

Whig Party a political group supported by people who opposed Andrew Jackson

Panic of 1837 a financial crisis that led to a severe economic depression

William Henry Harrison a general and the Whig presidential candidate in 1840

Academic Vocabulary

criteria basic requirements

Section Summary

SECTIONAL DIFFERENCES INCREASE

In Andrew Jackson's presidency, people's reactions to policies were based on where they lived and the economy of their region. The North's economy depended on trade and manufacturing. The North supported tariffs, which protected its industries. Southerners marketed crops to foreign countries and most opposed tariffs, which led to higher prices in manufactured items that they bought. Westerners wanted cheap land.

> Why did northerners disagree with southerners on the issue of tariffs?
>
> _____
>
> _____

THE TARIFF OF ABOMINATIONS

Northerners continued to demand high tariffs to guard their new industries from foreign competition. In 1828 Congress passed a law that southerners called the **Tariff of Abominations**. (An abomination is a hateful thing.) The tariff intensified sectional differences.

> How did the Tariff of Abominations help industries in the North?
> _____
> _____

THE STATES' RIGHTS DEBATE

Vice President John C. Calhoun of South Carolina argued that certain tariffs violated **states' rights**. The debate over states' rights led to the **nullification crisis**. Jackson opposed nullification. Calhoun resigned from office. South Carolina's legislature declared that a new 1832 tariff would not be collected in the state. **Daniel Webster** backed a unified nation. Congress finally agreed to lower the tariffs gradually. South Carolina's leaders agreed to obey the law but still backed the nullification idea.

> What caused the nullification crisis?
> _____
> _____

JACKSON ATTACKS THE BANK

President Jackson and many southern states questioned the constitutional legality of the Second Bank of the United States. However, in the case *McCulloch* v. *Maryland*, the Bank was found to be constitutional. Jackson moved most of the Bank's funds to state banks. This action caused inflation.

> What happened when the federal bank's funds were moved to state banks?
> _____
> _____

VAN BUREN'S PRESIDENCY

The **Whig Party** backed four candidates for president in 1836, and the Democrat, Martin Van Buren, won. When the country experienced the **Panic of 1837**, Van Buren was blamed. In 1840 the Whigs nominated **William Henry Harrison**, who won with an electoral landslide.

> Why might voters have chosen Harrison over Van Buren?
> _____
> _____

CHALLENGE ACTIVITY

Critical Thinking: Summarizing Design a poster that illustrates President Jackson's actions in his two terms. Use captions.
HSS ANALYSIS SKILL CS2

Chapter 9 The Age of Jackson

MAIN IDEAS

1. The Indian Removal Act authorized the relocation of Native Americans to the West.

2. Cherokee resistance to removal led to disagreement between Jackson and the Supreme Court.

3. Other Native Americans resisted removal with force.

 HSS 8.8

Students analyze the divergent paths of the American people in the West from 1800 to the mid-1800s and the challenges they faced.

Key Terms and People

Indian Removal Act the act that authorized the removal of Native Americans who lived east of the Mississippi River

Indian Territory the new homeland for Native Americans, which contained most of present-day Oklahoma

Bureau of Indian Affairs an agency created to oversee the federal policy toward Native Americans

Sequoya a Cherokee who used 86 characters to represent Cherokee syllables to create a written language

Worcester v. Georgia a case in which the U.S. Supreme Court ruled that the state of Georgia had no authority over the Cherokee

Trail of Tears an 800-mile forced march westward in which one-fourth of the 18,000 Cherokee died

Black Hawk a Sauk chief who decided to fight rather than be removed

Osceola Seminole leader who called on Native Americans to resist removal by force

Academic Vocabulary

contemporary existing at the same time

Section Summary

THE INDIAN REMOVAL ACT

President Jackson's policies toward American Indians were controversial. They had long lived in settlements from Georgia to Mississippi. Jackson and other politicians wanted this land for American farmers. Jackson pressured Congress to pass the

> Why were Jackson's policies toward Native Americans controversial?
>
> _____
>
> _____

Indian Removal Act in 1830. The **Indian Territory** was set aside as a new home for Native Americans.

The **Bureau of Indian Affairs** was established. Indian peoples began to be removed to Indian Territory. They lost their lands east of the Mississippi. On their trips to Indian Territory, many Native Americans died of cold, disease, and starvation. The Cherokee adopted some of the white culture to avoid conflicts. **Sequoya** helped the Cherokee create their own written language.

The Cherokee sued the state when the Georgia militia tried to remove them. In the case ***Worcester v. Georgia***, the U.S. Supreme Court ruled in favor of the Cherokee. Georgia ignored the ruling and removed the Cherokee. On the **Trail of Tears**, the Cherokee suffered from heat, cold, and exposure.

> **Why was the Indian Territory established?**
> _____
> _____

AMERICAN INDIAN RESISTANCE

Conflicts broke out in Illinois and Florida when Native Americans resisted removal with force. Chief **Black Hawk** led the Sauk of Illinois in raiding settlements and fighting the U.S. Army. The U.S. Army attacked the Sauk as they retreated, and the uprising ended. By 1850 American Indians had been driven from the Illinois region.

In Florida the Seminole also resisted removal. In 1832 some Seminole leaders were forced to sign a treaty that said they would withdraw from Florida in seven years. Any Seminole of African ancestry would be called a runaway slave. The Seminoles ignored the treaty. **Osceola** led his followers in the Second Seminole War. The Seminole won many battles. Some 1,500 U.S. soldiers died. After spending millions of dollars, U.S. officials gave up.

> **How did the Sauk resist removal?**
> _____
> _____

> **How did the outcome for the Seminole differ from that of other Native Americans?**
> _____
> _____

CHALLENGE ACTIVITY

Critical Thinking: Analyzing Write an essay explaining how your view of the Indian Removal Act would compare or contrast with the view of an easterner who wanted to settle on Native American lands.

HSS Skills Analysis

Chapter 10 Expanding West

HISTORY–SOCIAL SCIENCE STANDARDS

HSS 8.8 Students analyze the divergent paths of the American people in the West from 1800 to the mid-1800s and the challenges they faced.

HSS 8.9 Students analyze the early and steady attempts to abolish slavery and to realize the ideals of the Declaration of Independence.

HSS Analysis Skill CS 3 Students use a variety of maps and documents to identify physical and cultural features of neighborhoods, cities, states, and countries.

Date	Event
1811	John Jacob Astor founds Astoria
1818	
1819	Adams-Onís Treaty
1821	Agustin de Iturbide wins Mexican freedom
1830	Joseph Smith founds Church of Jesus Christ of the Latter-day Saints
1835	Texas Revolution begins in Gonzales
1836	
1839	John Sutter starts his colony in California
1844	James K. Polk elected president of the United States
1846	Treaty with British gives U.S. all Oregon land south of 49th parallel
1846	
1847	Brigham Young founds Salt Lake City
1850	
1853	
1869	Transcontinental railroad completed

COMPREHENSION AND CRITICAL THINKING

Use the answers to the following questions to fill in the graphic organizer.

1. **Recall and Identify** Add these events to the graphic organizer. Convention of 1818; Gadsden Purchase; California statehood; Fall of the Alamo; Bear Flag Revolt.

2. **Sequence** How many years passed between the first gathering of the Mormons and their final settlement in Utah?

_____

3. **Evaluate** Some would say that the events listed above demonstrate the validity of manifest destiny. Do you agree? Why or why not?

_____

_____

MAIN IDEAS

1. During the early 1800s, many Americans moved west of the Rocky Mountains to settle and trade.

2. The Mormons traveled West in search of religious freedom.

 HSS 8.8

Students analyze the divergent paths of the American people in the West from 1800 to the mid-1800s and the challenges they faced.

Key Terms and People

John Jacob Astor owner of the American Fur Company who founded the first important settlement in Oregon Country in 1811

mountain men fur traders and trappers who traveled to the Rocky Mountains and the Pacific Northwest in the early 1800s

Oregon Trail the main route from the Mississippi River to the West Coast in the early 1800s

Santa Fe Trail the route from Independence, Missouri, to Santa Fe, New Mexico

Mormons members of a religious group, formally known as the Church of Jesus Christ of Latter-day Saints, that moved west during the 1830s and 1840s

Brigham Young Mormon leader who founded Salt Lake City in 1847

Section Summary

AMERICANS MOVE WEST

In the early 1800s trappers and traders known as **mountain men** worked to supply the eastern fashion for fur hats and clothing. **John Jacob Astor**, owner of the American Fur Company, sent mountain men to the Pacific Northwest region that became known as Oregon Country. At this time, Oregon Country was inhabited by Native Americans. However, it was claimed by Russia, Spain, Great Britain, and the United States.

In 1811 Astor founded Astoria, which was the first major non-Native American settlement in the region, at the mouth of the Columbia River. After a series of treaties, Oregon Country soon became jointly occupied by Great Britain and the United

> What river provided the route from the Pacific to the interior of Oregon Country?
>
> _____
>
> _____

States. Many Americans began to move to the region, most of them following a challenging and dangerous route that became known as the **Oregon Trail**. It was common for families to band together and undertake the perilous six-month journey in wagon trains.

Another well-traveled route west, the **Santa Fe Trail**, was used mainly by traders. They loaded wagon trains with cloth and other manufactured goods that could be traded for horses, mules, and silver in the Mexican settlement of Santa Fe.

> What do you think was the main language spoken in Santa Fe at this time?
>
> _____
>
> _____

MORMONS TRAVEL WEST

One large group of settlers traveled west in search of religious freedom. Joseph Smith founded the Church of Jesus Christ of Latter-day Saints in 1830 in western New York state. Although church membership grew rapidly, the converts, known as **Mormons**, were dogged by local hostility. To protect his growing community from persecution, Smith led his followers to a series of settlements in Ohio, Missouri, and Illinois.

> Do some research on the Mormons. Trace the path that they took from New York to Salt Lake City.

> Using the library or an online resource, find out how many members worldwide the Mormon Church has today.
>
> _____
>
> _____

When Smith was murdered by an anti-Mormon mob in 1844, **Brigham Young** led the Mormons to a desert valley near the Great Salt Lake in what is now Utah. There the Mormons planned and built Salt Lake City and settled in the area. By December 1860 the Mormon population of Utah stood at about 40,000.

CHALLENGE ACTIVITY

Critical Thinking: Drawing Inferences Make a list of supplies that a family of four would need to make a six-month journey by wagon train through the American West during the 1830s. **HSS Analysis Skills HI 1, HI 2, HI 4, HI 6**

Chapter 10 Expanding West

MAIN IDEAS

1. Many American settlers moved to Texas after Mexico achieved independence from Spain.

2. Texans revolted from Mexican rule and established an independent nation.

 HSS 8.8

Students analyze the divergent paths of the American people in the West from 1800 to the mid-1800s and the challenges they faced.

Key Terms and People

Father Miguel Hidalgo y Costilla priest who led the first major Mexican revolt against Spanish rule in 1810

empresarios agents of the Mexican republic hired to bring settlers to Texas

Stephen F. Austin empresario who established the first American colony in Texas

Antonio López de Santa Anna Mexican leader who came to power in 1830 and suspended Mexico's constitution

Alamo an old mission in San Antonio occupied by Texan revolutionary forces in 1836

Battle of San Jacinto decisive victory that gave Texas independence from Mexico

Academic Vocabulary

explicit fully revealed without vagueness

Section Summary

AMERICAN SETTLERS MOVE TO TEXAS

In the early 1800s, the region we now know as the American Southwest was part of Mexico, which in turn was part of the vast Spanish empire in the Americas. Mexico struggled against Spanish rule. A revolt led by **Father Miguel Hidalgo y Costilla** in 1810 failed, but the rebellion he started grew. In 1821 Mexico became independent.

In order to establish control of Texas, the new Mexican republic hired agents known as **empresarios** to bring settlers there. One of these, **Stephen F. Austin**, selected a site on the lower Colorado River and settled 300 families, mostly from the southern

> Use an online browser or another resource to research "the Old 300."

states. These settlers often ignored Mexican laws, including Mexico's law forbidding slavery.

Tension grew between Mexico's central government and the American settlers. Colonists were angry when **Antonio López de Santa Anna** came to power in 1830 and suspended Mexico's constitution. Austin was imprisoned for a year and a half. When he returned to Texas, he began urging Texans to rebel against Mexico.

> From what region of the United States did most settlers come to Texas?
> _____
> _____

> Underline the sentence that helps explain why tension grew between the central Mexican government and the American settlers in Texas.

TEXANS REVOLT AGAINST MEXICO

Hostilities began with a battle at Gonzalez in 1835. Santa Anna inflicted two brutal defeats on the Texans at the **Alamo** and Goliad. Within a month, however, Texas forces under Sam Houston had won a decisive victory over Santa Anna at the **Battle of San Jacinto**. Santa Anna signed a treaty giving Texas its independence.

> Use the library or an online resource to find an account of the famous siege of the Alamo.

Most people in the new Republic of Texas hoped that Texas would join the United States. However, U.S. President Andrew Jackson was concerned about two factors. He was worried that admitting Texas as a slave state would upset the fragile balance between free and slave states in the Union. Also, Jackson feared that annexing Texas might lead to a war with Mexico.

As the annexation of Texas was delayed, more American settlers came from nearby southern states, often bringing slaves with them to work the land and to grow cotton. Tensions between Mexico and Texas remained high. After a few unsettled years, Texas President Sam Houston signed a peace treaty with Mexico in 1844.

CHALLENGE ACTIVITY

Critical Thinking: Evaluation Take sides in a debate as to whether Texas should join the United States or remain an independent nation. Write a defense of your position. **HSS Analysis Skills HR 5, HI 1, HI 2, HI 3, HI 6**

Chapter 10 Expanding West

MAIN IDEAS

1. Many Americans believed that the nation had a manifest destiny to claim new lands in the West.

2. As a result of the Mexican-American War, the United States added territory in the Southwest.

3. American settlement in the Mexican Cession produced conflict and a blending of cultures.

 HSS 8.8

Students analyze the divergent paths of the American people in the West from 1800 to the mid-1800s and the challenges they faced.

Key Terms and People

manifest destiny belief that America's fate was to conquer land all the way to the Pacific Ocean

James K. Polk U.S. president, elected in 1844, whose administration annexed both Texas and Oregon

vaqueros cowboys who managed the large herds of cattle and sheep owned by wealthy California settlers

Californios Spanish colonists and their descendants living in California

Bear Flag Revolt rebellion of American settlers against the Californios in 1846

Treaty of Guadalupe Hidalgo 1848 peace treaty between Mexico and the United States

Gadsden Purchase purchase from Mexico of the southern parts of present-day New Mexico and Arizona in 1853

Academic Vocabulary

elements the basic parts of an individual's surroundings

Section Summary

MANIFEST DESTINY

The idea of **manifest destiny** loomed large in the election of 1844. The new president, **James K. Polk**, promised to annex both Texas and Oregon. In a treaty with Britain in 1846 the United States gained all Oregon land south of the 49th parallel. This completed the present-day border between the United States and Canada. In 1845 the United States annexed Texas.

> Use the library or an online resource for an understanding of why the idea of manifest destiny may have been so attractive during the 1840s.

Since the early 1700s, scattered settlements of Spanish colonists in the present-day Southwest had been engaged in farming and ranching, often using Native American labor. In California many of these settlers, known as **Californios**, profited when Mexico won independence from Spain. American settlers began coming to California and set up cattle and sheep ranches managed by *vaqueros.*

> Do you think the Californios resented the arrival of large numbers of American settlers? Why or why not?
>
> _____
> _____
> _____
> _____

MEXICAN-AMERICAN WAR

Since the Texas Revolution, the border between Mexico and Texas had been in dispute. Mexico claimed the border lay along the Nueces River while the United States claimed the Rio Grande as the border. In 1845 President Polk sent troops to the Rio Grande. When Mexican soldiers attacked them, Congress declared war on Mexico. U.S. troops pushed into Mexico, winning victories and finally capturing Mexico City. A successful revolt against the Californios near Sonoma, known as the **Bear Flag Revolt**, proclaimed the independent Republic of California.

> Some Americans at the time thought President Polk provoked the Mexican attack by stationing soldiers on the Rio Grande. Do you agree? Explain your answer.
>
> _____
> _____

> Use the library or an online resource to find a map showing the territorial growth of the United States during this period.

AMERICAN SETTLEMENT IN THE MEXICAN CESSION

The **Treaty of Guadalupe Hidalgo**, which ended the war in 1848, increased the land area of the United States by almost 25 percent. In 1853 the **Gadsden Purchase** fixed the continental boundaries of the United States.

As American settlers flooded the Southwest, some cultural encounters led to conflict. New settlers often ignored Mexican legal ideas, such as community property or community water rights. However, mutually beneficial trade patterns emerged.

> Why do you think the issue of water rights is much more serious in the West than it is in the East?
>
> _____
> _____
> _____

CHALLENGE ACTIVITY

Critical Thinking: Identify Cause and Effect Write a law regulating water rights. **HSS Analysis Skills CS 3, HI 1, HI 2, HI 4, HI 6**

Chapter 10 Expanding West

MAIN IDEAS
1. The discovery of gold brought settlers to California.
2. The gold rush had a lasting impact on California's population and economy.

 HSS 8.8
Students analyze the divergent paths of the American people in the West from 1800 to the mid-1800s and the challenges they faced.

Key Terms and People

John Sutter Swiss immigrant who started the first Anglo-Californian colony in 1839

Donner party a group of western travelers who were trapped crossing the Sierra Nevada in the winter of 1846-47

forty-niners gold-seeking migrants who traveled to California in 1849

prospect search for gold

placer miners miners who used pans or other devices to wash gold nuggets from loose rock or gravel

Section Summary

DISCOVERY OF GOLD BRINGS SETTLERS

Before 1840 few Americans settled in California, although there was a considerable trade there between merchants from Mexico and the United States. However, after Mexico allowed **John Sutter** to establish a colony in 1839, American settlers began arriving in greater numbers. In a tragic incident, heavy snows in the Sierra Nevada trapped a group of travelers known as the **Donner party**. Half of the travelers either froze or starved to death.

When gold was discovered at Sutter's Mill in 1848, the news spread across the country. During 1849 about 80,000 **forty-niners** came to California hoping to strike it rich. Most of them arrived in the small port town of San Francisco. Within a year, the population of San Francisco grew from around 800 to more than 25,000.

> **What country was California part of in 1840?**
> _____
> _____

Mining methods varied by the time of year and the location of the claim. **Placer miners** would **prospect** by using pans or other devices to wash gold nuggets out of the loose rock and gravel. Richer miners established companies to dig shafts and tunnels. Many individual success stories inspired prospectors. However, the good luck that made some miners wealthy never came to thousands of gold seekers. Most of them found little except misery and debt.

Mining camps sprang up wherever enough people gathered to look for gold. Among the gold seekers were thousands of immigrants from Mexico, China, Europe, and South America. Many found that they could earn a living by supplying miners with basic services like cooking, washing clothes, operating boardinghouses, or even providing legal services. Biddy Mason and her family, slaves brought to California by a forty-niner from Georgia, gained their freedom and managed to buy some land near the village of Los Angeles. Soon Mason became one of the weathiest landowners in California.

> Do you think mining experience would have helped the average forty-niner? Explain your answer.
> _____
> _____
> _____
> _____

> Who had more "job security," the miners or the service providers? Why?
> _____
> _____
> _____
> _____

IMPACT ON CALIFORNIA

The forty-niners brought a population explosion and an economic boom to California. It became the 31st state of the Union in 1850. As the gold rush faded, many Californians took to farming and ranching. However, California remained isolated from the rest of the country until the transcontinental railroad was completed in 1869.

> What consequence of the gold rush made California eligible for statehood?
> _____
> _____

CHALLENGE ACTIVITY

Critical Thinking: Drawing Inferences Design and write a brochure inviting easterners to come to Sutter's colony in California and start a new life. **HSS Analysis Skills CS 3, HR 2, HR 3, HR 5, HI 4**

Chapter 11 The North

> **HISTORY–SOCIAL SCIENCE STANDARDS**
> **HSS 8.6** Students analyze the divergent paths of the American people from 1800 to the mid-1800s and the challenges they faced, with emphasis on the Northeast.
> **HSS Analysis Skill HI 1** Students explain the central issues and problems from the past.
> **HSS Analysis Skill HI 2** Students understand and distinguish cause, effect, sequence, and correlation in historical events, including the long- and short-term causal relations.

CHAPTER SUMMARY

The Industrial Revolution		The Transportation Revolution
Caused by		Caused by
↓		↓
invention of machines		invention of the railroad
_____		_____
led to		led to
↓		↓
_____		_____

COMPREHENSION AND CRITICAL THINKING

Use the answers to the following questions to fill in the graphic organizer above.

1. **Comparing** Compare one cause of the Industrial Revolution with one cause of the Transportation Revolution.

   _____

2. **Identify Cause and Effect** Name two ways in which the Industrial Revolution changed the lives of the American people.

   _____

3. **Identify Cause and Effect** Name two ways in which the Transportation Revolution changed the lives of the American people.

   _____

Chapter 11 The North

MAIN IDEAS

1. The invention of new machines in Great Britain led to the beginning of the Industrial Revolution.

2. The development of new machines and processes brought the Industrial Revolution to the United States.

3. Despite a slow start in manufacturing, the United States made rapid improvements during the War of 1812.

 HSS 8.6

Students analyze the divergent paths of the American people from 1800 to the mid-1800s and the challenges they faced, with emphasis on the Northeast.

Key Terms and People

Industrial Revolution a period of rapid growth in the use of machines in manufacturing and production

textiles cloth items

Richard Arkwright an inventor who patented a large spinning machine, called the water frame, that ran on water power and created dozens of cotton threads at once

Samuel Slater a skilled British mechanic who could build the new textile machines

technology the tools used to produce items or to do work

Eli Whitney an inventor with an idea for mass-producing guns

interchangeable parts pieces that are exactly the same

mass production the efficient production of large numbers of identical goods

Academic Vocabulary

efficient productive and not wasteful

Section Summary

THE INDUSTRIAL REVOLUTION

In the early 1700s, most people in the United States and Europe made a living by farming. Female family members often used hand tools to make cloth for families. The sale of extra cloth earned money. Skilled workers such as blacksmiths set up shops to earn money by manufacturing goods by hand. The **Industrial Revolution** would completely change that way of life. By the mid-1700s, cities and popu-

> In what way were goods made in the early 1700s?
>
> _____
> _____
> _____

lations had grown. Demand increased for ways to make items faster and more efficiently.

Textiles provided the first breakthrough. **Richard Arkwright** invented a machine that lowered the cost of cotton cloth and raised production speed. Most textile mills using such machines were built near streams to use running water for power.

> **In what way did Arkwright's machine make history?**
> _____
> _____

SLATER AND HIS SECRETS

Samuel Slater knew how to build machines that were used in Britain to make cloth more efficiently. He emigrated to the United States, and with Moses Brown opened a mill in Pawtucket, Rhode Island. The mill made cotton thread by machine. It was a success. Most mills were in the northeast, the region with many rivers and streams for power.

> **What was Slater's secret?**
> _____
> _____

A MANUFACURING BREAKTHROUGH

In the 1790s, U.S. gun makers could not produce muskets quickly enough if there was a war. Better **technology** was needed. Eli Whitney had the idea of manufacturing using **interchangeable parts**. Whitney assembled muskets for President Adams. His idea worked. **Mass production** was soon used in factories making interchangeable parts.

> **What was Whitney's revolutionary idea?**
> _____
> _____
> _____

MANUFACTURING'S SLOW START

U.S. manufacturing spread slowly. People who could buy good farmland would not work for low factory wages. British goods were cheaper than American goods. However, during the War of 1812 many Americans learned that they had relied on foreign goods too much. In 1815 the war ended and free trade returned. Businesspeople wanted to lead the nation into a time of industrial growth.

> **Why had Americans relied on foreign goods too much?**
> _____
> _____

CHALLENGE ACTIVITY

Critical Thinking: Rating In comparing the three inventors in Section 1, rate them from 1 to 3. Defend your rating order in a brief essay.
HSS Analysis Skills CR 4.

Section 2

MAIN IDEAS

1. The spread of mills in the Northeast changed workers' lives.

2. The Lowell system revolutionized the textile industry in the Northeast.

3. Workers organized to reform working conditions

 HSS 8.6

Students analyze the divergent paths of the American people from 1800 to the mid-1800s and the challenges they faced, with emphasis on the Northeast.

Key Terms and People

Rhode Island system Samuel Slater's strategy of hiring families and dividing factory work into simple tasks

Francis Cabot Lowell a New England businessman who built a loom that could both weave thread and spin cloth in the same mill

Lowell system Lowell's practice of hiring young unmarried women to work in his mills

trade unions groups of skilled workers that tried to improve members' pay and working conditions

strikes union workers' refusal to work until their employers met their demands

Sarah G. Bagley a mill worker who founded the Lowell Female Labor Reform Association

Academic Vocabulary

concrete specific, real

Section Summary

FACTORY FAMILIES

Samuel Slater had difficulty in hiring enough people to work in his mills. Young male apprentices often left because their work was boring. Slater began hiring families who moved to Pawtucket. Children usually earned in one week what an adult was paid for one day's work. Slater constructed housing for the workers. He paid workers in credit at the company store rather than paying them money so that he could reinvest money in his business. Slater's method was known as the **Rhode Island system**. Many northeastern mill owners imitated Slater's system.

> How much did child workers earn in factories?
>
> _____

THE LOWELL SYSTEM

Francis Cabot Lowell developed a different approach called the **Lowell system**. It transformed the Northeast's textile industry. With the aid of a company, Lowell built mills in Waltham and Lowell, both in Massachusetts. The factories were clean and the workers' boardinghouses were neat. Many young women, called Lowell girls, journeyed from across New England to earn money instead of earning nothing on the family farm. The Lowell girls were encouraged to take classes and join clubs. However, they worked 12- to 14-hour days, and cotton dust caused health problems for them.

> **Name one advantage and one disadvantage of Lowell mill work.**
> _____
> _____
> _____

WORKERS ORGANIZE

Factory workers' wages went down as people competed for jobs. Immigrants also competed for jobs. The Panic of 1837 led to unemployment for many. Skilled workers started **trade unions** for protection. Sometimes union members held **strikes**. But most strikes were not very successful.

> **Why did workers' pay decrease?**
> _____
> _____

LABOR REFORM EFFORTS

Sarah G. Bagley battled for the workers. She was the first highly ranked woman in America's labor movement. In 1840 President Martin Van Buren had given a 10-hour workday to many federal employees. Bagley supported the 10-hour workday for all workers. The Unions won some legal victories. Some states passed 10-hour workday laws. But companies often found ways to get around them. Other states did not pass the 10-hour workday laws. Union supporters kept fighting for improved working conditions during the 1800s.

> **What did workers achieve in the mid-1800s?**
> _____
> _____

CHALLENGE ACTIVITY

Critical Thinking: Contrasting Write a letter to the editor contrasting the lives of workers in Slater's mills and Lowell's mills.
HSS Analysis Skills CS 1.

Chapter 11 The North

> **MAIN IDEAS**
> 1. The Transportation Revolution affected trade and daily life.
> 2. The steamboat was one of the first developments of the Transportation Revolution.
> 3. Railroads were a vital part of the Transportation Revolution.
> 4. The Transportation Revolution brought many changes to American life and industry.

 HSS 8.6
Students analyze the divergent paths of the American people from 1800 to the mid-1800s and the challenges they faced, with emphasis on the Northeast.

Key Terms and People

Transportation Revolution rapid growth in the speed and convenience of travel
Robert Fulton an American who had the first full-time commercial steamboat in the United States
Clermont a steamboat that traveled up the Hudson River with no trouble
Gibbons v. *Ogden* the first U.S. Supreme Court ruling on commerce between states
Peter Cooper built the *Tom Thumb,* a small steam train with great power and speed

Section Summary

NEW WAYS TO TRAVEL

The **Transportation Revolution** changed life in the 1800s, along with the Industrial Revolution, by speeding travel and decreasing time and cost of shipping goods between the East and the West. People and information began traveling at much higher speeds. New towns and businesses sprang up with improved communication, travel, and trade. The steamboat and the railroad, new kinds of transportation, quickened the pace of American life.

> Why did information begin traveling at higher speeds?
> _____
> _____
> _____

> In what way might the pace of American life have increased?
> _____
> _____
> _____

THE STEAMBOAT

In the late 1700s American and European inventors advanced steam-powered boats. **Robert Fulton** tested the *Clermont* in the United States. The successful test helped launch the steamboat era.

Steamboats cut months off the time to travel from one place to another. They made trips up rivers cheaper and easier. Shipping goods from East to West, West to East, or overseas also was easier.

Sometimes the changes in transportation led to legal conflicts. In a landmark case, *Gibbons* v. *Ogden*, the court ruled that federal shipping laws overruled state shipping laws.

> **How did steamboats affect shipping?**
> _____
> _____
> _____

AMERICAN RAILROADS

About 1830 **Peter Cooper** built the *Tom Thumb,* a small but fast train. Excitement over rail travel grew in the mid-1800s. By 1860 about 30,000 miles of railroad tracks joined nearly every major eastern U.S. city. Trains took goods to faraway markets. Train travel averaged about 20 miles per hour and could be dangerous because of fires and derailment. But the dangers did not put off travelers who wanted to go places faster.

> **Circle the sentence that explains why travelers put up with the dangers of railroad travel.**

POWERING THE TRANSPORTATION REVOLUTION

Construction crews put down track through rock and over mountains and rivers. Trains brought new residents and raw materials for industry to cities, spurring growth. As faster locomotives were built, coal replaced wood as a source of fuel because of its greater efficiency. That led to growth in the mining industry. Steel was used for railroad tracks, so the demand for steel increased. Railroad transportation also helped logging expand because wood was needed to build new houses in the growing cities. Chicago, on Lake Michigan, became a hub for national transportation.

> **What helped the steel industry?**
> _____
> _____

CHALLENGE ACTIVITY

Critical Thinking: Designing Design a four-page brochure advertising the wonders of travel by steamboat or train. **HSS Analysis Skills HI 1**

HSS 8.6
Students analyze the divergent paths of the American people from 1800 to the mid-1800s and the challenges they faced, with emphasis on the Northeast.

MAIN IDEAS
1. The telegraph made swift communication possible from coast to coast.
2. With the shift to steam power, businesses built factories closer to cities and transportation centers.
3. Improved farm equipment and other labor-saving devices made life easier for many Americans.
4. New inventions changed lives in American homes.

Key Terms and People

Samuel F. B. Morse the inventor of the telegraph

telegraph a device that could send information over wires across great distances

Morse code a system in which dots and dashes are used to stand for each letter of the alphabet

John Deere a blacksmith who first used the steel plow design

Cyrus McCormick developer of a new harvesting machine called a mechanical reaper

Isaac Singer made improvements in the design of the sewing machine

Section Summary

MESSAGES BY WIRE

Samuel F. B. Morse invented the **telegraph** in 1832. Morse used the work of two other scientists in making this practical machine. In a telegraph, pulses, or surges, of electric current were carried over wire. The operator touched a bar, called a telegraph key, that set the length of each pulse. At the wire's other end, the pulses were changed into clicks. A short click was a dot. A dash was a long click. Morse's assistant, Alfred Lewis Vail, developed the **Morse code**. Some people did not think Morse could actually read messages sent across long distances. But during the 1844 Democratic Convention in Baltimore, Maryland, a telegraph wired news of a nomination to politicians in Washington. Soon

What was the invention of the telegraph based on?

_____

_____

Did everyone accept the telegraph's power at first?

_____

_____

telegraphs were relaying messages for businesses, the government, newspapers, and private citizens. Telegraph lines were strung on poles next to railroad tracks across the country.

> **Why did telegraph messages become so widely used?**
> _____
> _____
> _____

NEW FACTORIES

Most factories, operating on water power at first, had to be built near water. With the use of steam engines, factories could be built almost anywhere. Still, most were in the Northeast. By 1860 New England had as many factories as all of the South had. Many new factories were near cities and transportation centers, giving them better access to workers. In addition, by the 1840s new machinery could produce interchangeable parts.

> **Why could newer factories be built almost anywhere?**
> _____
> _____
> _____

BETTER FARM EQUIPMENT

John Deere was selling 1,000 steel plows a year by 1846. **Cyrus McCormick** mass-produced his reapers in a large Chicago factory. His company advertised, provided service, and let customers buy on credit. The plow and the reaper allowed Midwestern farmers to harvest huge wheat fields.

> **Circle the sentence that explains what new methods McCormick used to persuade people to buy his reapers.**

CHANGING LIFE AT HOME

The sewing machine was among the American inventions that made home life easier. **Isaac Singer** modified the sewing machine, and worked hard to sell his product. Prices of many household items had decreased, giving many more people the ability to afford them. Many more cities built public water systems, although very few homes installed plumbing above the first floor.

> **Why did the purchase of many household items increase?**
> _____
> _____
> _____

CHALLENGE ACTIVITY

Critical Thinking: Evaluating Write a brief essay explaining which invention mentioned made the biggest change in people's lives.
HSS Analysis Skills HI 2, CR 3

Chapter 12 The South

HISTORY–SOCIAL SCIENCE STANDARDS
HSS 8.7 Students analyze the divergent paths of the American people in the South from 1800 to the mid-1800s and the challenges they faced.
HSS 8.9 Students analyze the early and steady attempts to abolish slavery and to realize the ideals of the Declaration of Independence.

CHAPTER SUMMARY

CAUSE		EFFECT
The price of cash crops drops.	⟶	The need for slaves decreases.
Eli Whitney's cotton gin makes it profitable to grow cotton.	⟶	
	⟶	Most Southerners owned few or no slaves.
Slave behavior was strictly controlled.	⟶	Slaves created their own culture, and some rebelled.

COMPREHENSION AND CRITICAL THINKING

Use the answers to the following questions to fill in the graphic organizer above.

1. Draw Conclusions How did Whitney's cotton gin affect the value of slaves?

_____

_____

2. Identify Cause and Effect Why did only one-third of southerners own slaves?

_____

_____

3. Make Judgments Do you think there was a time when southern farmers depended on cash crops? Why or why not?

_____

_____

Chapter 12 The South

Section 1

MAIN IDEAS

1. The invention of the cotton gin revived the economy of the South.

2. The cotton gin created a cotton boom in which farmers grew little else.

3. Some people encouraged the southerners to focus on other crops and industries.

 HSS 8.7

Students analyze the divergent paths of the American people in the South from 1800 to the mid-1800s and the challenges they faced.

Key Terms and People

cotton gin machine that separates cotton from its seeds

planters large-scale farmers who held more than 20 slaves

cotton belt nickname for the region that grew most of the country's cotton crop

factors crop brokers who arranged transportation of goods aboard trading ships

Tredegar Iron Works in its day, the only large southern factory that made iron products

Academic Vocabulary

primary main, most important

Section Summary

THE SOUTH'S AGRICULTURAL ECONOMY

After the American Revolution, the use of slaves began to decline. One reason was that many Americans felt keeping slaves in a nation founded on freedom was wrong. Another reason was that using slaves was no longer economically profitable to the owner. That is because crop prices fell. Farmers planted less, so they needed less slave labor. Some found it cheaper to just free their slaves.

ELI WHITNEY AND THE COTTON GIN

Cotton was not a new crop to the South. However, few farmers planted much, for the short-staple cotton that grew well there was very hard to separate

> Why was it difficult to harvest cotton before the invention of the cotton gin?
>
> _____
>
> _____

from its seeds. Northerner Eli Whitney changed
that when he invented the **cotton gin**. This hand-
cranked cylinder easily pulled cotton and seeds
apart. Cotton became profitable. A cotton gin could
clean as much cotton as planters could plant and
their slaves could pick. A **planter** was a large-scale
farmer who held more than 20 slaves.

THE COTTON BOOM

For southern farmers, cotton had many advantages
over other crops. Unlike food products, cotton
could be stored for long periods of time. Plus its
lightness made it fairly inexpensive to transport. As
a result, the cotton-supported slave trade grew, even
as Congress worked to limit slavery in the nation.

> **What were the advantages of cotton compared to other crops?**
> _____
> _____

THE COTTON TRADE

Most of the country's cotton was produced in the
cotton belt, which stretched from South Carolina
to Texas. Because of a lack of roads and canals,
southern farmers relied on rivers to move their
crops. When their cotton reached port, farmers sold
it to merchants, who used brokers called **factors** to
arrange shipping.

> **Why were the region's rivers especially important to southern farmers?**
> _____
> _____

AGRICULTURAL DIVERSITY

Scientific agriculture, or the use of scientific methods to improve farming, encouraged southern
farmers to rotate the kinds of crops they planted.
So, farmers also grew corn, rice, sugarcane, wheat,
tobacco, hemp, and flax. Some industries, such as
the **Tredegar Iron Works**, also flourished, but most
of the South focused on farming.

> **Circle the definition of scientific agriculture.**

CHALLENGE ACTIVITY

Critical Thinking: Write to Explain What if a new fiber came along that
replaced cotton in clothing. What effect would this development have
on cotton planters? Write a paragraph explaining how falling cotton
 ̄ ̄ ̄ight lead to less demand for farm workers. **HSS Analysis Skills:**
II 5

Chapter 12 The South

MAIN IDEAS
1. Southern society and culture consisted of four main groups.
2. Free African Americans in the South faced a great deal of discrimination.

 HSS 8.7
Students analyze the divergent paths of the American people in the South from 1800 to the mid-1800s and the challenges they faced.

Key Terms and People

yeomen owners of small farms

Section Summary

FOUR MAIN GROUPS OF SOUTHERN SOCIETY

Only about one-third of all southerners owned slaves. Far fewer were actually wealthy planters. However, those few planters were among the most influential southern citizens, and many were political leaders.

On the vast plantations, the planter ran the farm business. A wealthy planter would have overseers to help him. The planter's wife ran the household, which included many house slaves. She also was in charge of important social events such as dances and dinners.

Most southern farmers were yeomen. **Yeomen** owned small farms averaging about 100 acres, and often they worked side by side with the few slaves they might own.

Many white southerners were poor. They owned no slaves at all. Often they lived on land that could not grow crops. These farmers were at the bottom of the economic ladder.

How was a yeoman different from a planter?

_____

_____

_____

RELIGION AND SOCIETY

Religion was central to southern life. One reason was its social impact. Often farm families only saw their far-spread neighbors at church functions. Some southerners also believed that Christianity

Why was religion central to southern life?

_____

_____

justified slavery—a belief not shared by Christians in the North.

SOUTHERN CITIES

The economy of the South also depended on the businesses conducted in its busy cities. As in northern cities, southern cities provided many services to residents, including water systems and street maintenance. Southern cities used slave labor, too. Businesses either owned slaves or hired them out from nearby planters.

> How were southern cities like northern cities?
> _____
> _____
> _____

FREE AFRICAN AMERICANS

Not all African Americans were slaves. Some were free. Some had been born free. Others had bought their freedom from their slaveowners or had run away. About half of these free African Americans lived in the South.

> Circle the sentence describing how many of the free African Americans lived in the South.

The presence of free African Americans concerned some white southerners. They worried that those who were free would incite those who were enslaved to rise up against their slaveowners. As a result, southern cities and states passed laws aimed at limiting the rights of these free African Americans. Virginia went so far as to forbid former slaves from living in the state without permission.

> According to many southerners, how did free African Americans threaten the South's slave system?
> _____
> _____
> _____

Free African Americans posed another threat to white southerners, too. Many whites felt the African American could not survive outside of slavery, and these free men and women proved that was wrong.

CHALLENGE ACTIVITY

Critical Thinking: Evaluate How effective were the laws limiting the rights of free African Americans? Write five questions that could be answered by historical study and online research. **HSS Analysis Skills: HR 1**

Chapter 12 The South

MAIN IDEAS

1. Slaves worked at a variety of jobs on plantations.
2. Life under slavery was difficult, and punishments severe.
3. Slave culture centered around family, community, and religion.
4. Slave uprisings led to stricter slave codes in many states.

 HSS 8.9
Students analyze the early and steady attempts to abolish slavery and to realize the ideals of the Declaration of Independence.

Key Terms and People

folktales stories with a moral

spirituals songs that combine African and European music, and express religious beliefs

Nat Turner Virginia slave who led a rebellion against slaveholders in 1831

Nat Turner's Rebellion the name given to the 1831 rebellion led by Nat Turner

Academic Vocabulary

aspect part

Section Summary

SLAVES AND WORK

Most planters used the gang labor system to get their fields farmed. In this system, enslaved men, women, and children over 10 years of age all worked the same fieldwork from dawn until dark.

Slaves with special skills often were rented out by their owners. Sometimes they were allowed to keep part of what they earned. Because of this, some skilled slaves saved enough money to buy their freedom.

> **How could skilled slaves buy their freedom?**
> _____
> _____
> _____

LIFE UNDER SLAVERY

To most southern slaveholders, slaves were property, not people. As property, slaves could be bought and sold. Usually, this business occurred at a slave auction. At these auctions, family members could be sold away from each other forever.

> **What could happen to family members at a slave auction?**
> _____
> _____

Many slaveholders used cruel punishments to make sure their slaves stayed obedient. In addition, many states passed strict slave codes. These laws limited what slaves could do. For example, in some states it was illegal to teach slaves to read and write.

> **How were laws used to control slaves?**
> _____
> _____
> _____

SLAVE CULTURE

For enslaved African Americans, the family was the most important part of their lives. Parents made sure their children knew the African part of their history, including African customs and traditions. Because they could not read and write, they passed on this information verbally. Some of their stories were **folktales**—stories with morals—to teach children how to survive slavery.

> **How did slaves keep alive their culture?**
> _____
> _____
> _____

Religion was also an important part of the culture of enslaved African Americans. Christian slaves believed that, in God's eyes, they were equal to anyone else. They held onto the hope that someday they would be freed. Often, these beliefs were expressed in the **spirituals** they sang.

CHALLENGING SLAVERY

Enslaved African Americans found a variety of ways to protest their treatment. Some even ran away. Most runaways, though, were forced to return. That was because getting all the way North to freedom was filled with dangers and hardships.

Sometimes, slaves protested with violence. They risked certain punishment. This was true of Virginia slave **Nat Turner**. During **Nat Turner's Rebellion** in 1831, slaves killed about 60 white people. In the end, though, more that 100 slaves were killed and Turner was executed. As a result, many states strengthened their slave codes.

> **When did Nat Turner's Rebellion occur?**
> _____

CHALLENGE ACTIVITY

Critical Thinking: Analyze Write a paragraph explaining why southern slaveholders would want to keep slaves from learning to read and write.
HSS Analysis Skills: HI 2, CR 2

Chapter 13 New Movements in America

HISTORY–SOCIAL SCIENCE STANDARDS

HSS 8.6 Students analyze the divergent paths of the American people from 1800 to the mid-1800s and the challenges they faced, with emphasis on the Northeast.
HSS 8.9 Students analyze the early and steady attempts to abolish slavery and to realize the ideals of the Declaration of Independence.

CHAPTER SUMMARY

	THEN . . .	Nativists start to worry about their job security.
The philosophy of transcendentalism becomes popular.	THEN . . .	Many American writers are influenced by these new beliefs.
The Second Great Awakening begins.	THEN . . .	
Many Americans believe slavery should end.	THEN . . .	The Underground Railroad helps slaves escape from the South.
	THEN . . .	

COMPREHENSION AND CRITICAL THINKING

Use the answers to the following questions to fill in the graphic organizer above.

1. Explain What made nativists worry about their job security?

_____

_____

2. Identify Sequence What reforms followed the Second Great Awakening?

_____

_____

3. Draw a Conclusion What helped start the women's rights movement?

_____

_____

Chapter 13 New Movements in America

MAIN IDEAS

1. Millions of immigrants, mostly German and Irish, arrived in the United States despite anti-immigrant movements.

2. Industrialization led to the growth of cities.

3. American cities experienced urban problems due to rapid growth.

 HSS 8.6

Students analyze the divergent paths of the American people from 1800 to the mid-1800s and the challenges they faced, with emphasis on the Northeast.

Key Terms and People

nativists people who opposed immigration

Know-Nothing Party political organization founded by nativists in 1849

middle class a social and economic level between the wealthy and the poor

tenements dirty, unsafe housing structures in which the cities' poor were forced to live

Academic Vocabulary

implicit understood though not clearly put into words

Section Summary

MILLIONS OF IMMIGRANTS ARRIVE

Between 1840 and 1860, more than 4 million immigrants came to the United States. Many came from Ireland, fleeing starvation that came with a terrible potato famine there. The famine also meant that many Irish immigrants arrived poor. These immigrants often got jobs working long hours for little pay.

Unlike the Irish, immigrants from Germany often arrived with some money. Many came to America after a revolution in their homeland. Others came for the opportunities America offered. Many bought farmland in America's Midwest. Others settled and worked in cities.

> Compare and contrast Irish and German immigration between 1840 and 1860.
>
> _____
> _____
> _____

THE NATIVIST RESPONSE

To many native-born Americans, the new immi-

> Why did increasing numbers of immigrants make nativists worry about their jobs?
>
> _____
> _____

grants posed a threat. Americans worried that
immigrants would take away their jobs. Immigrants
would do the same work but for less money. The
Americans also mistrusted immigrants who were
Catholic. In Europe, Protestants and Catholics had
a history of conflicts. Americans who opposed
immigration for these reasons were known as **nativists**. Together, the nativists formed a political group
called the **Know-Nothing Party** to try limiting
immigration.

THE GROWTH OF CITIES

In the mid-1800s, the Industrial Revolution encouraged rapid growth in America's cities. The jobs the
Industrial Revolution created also helped build a
middle class—a social and economic level between
the wealthy and the poor. These new urban dwell-
ers enjoyed the culture in America's cities. Libraries,
clubs, and theaters grew as the cities grew.

> Why do you think culture changed after the Industrial Revolution?
>
> _____
>
> _____

URBAN PROBLEMS

The people who moved to the city to work could
afford only tenement rents. **Tenements** were poorly
designed housing structures that were dirty, overcrowded, and unsafe. Cities had not yet learned how
to deal with the filth and garbage generated by so
many people, and killer epidemics resulted. Crime
and fires also plagued the fast-growing cities of the
United States.

> Why were living conditions so poor in urban areas?
>
> _____
>
> _____

CHALLENGE ACTIVITY

Critical Thinking: Summarize List some of the health and safety issues
that plagued America's cities during the first half of the 1800s.
ISS Analysis Skills HI 1, HI 2

Chapter 13 New Movements in America

Section 2

MAIN IDEAS

1. Transcendentalists and utopian communities withdrew from American society.

2. American Romantic painters and writers made important contributions to art and literature.

 HSS 8.6
Students analyze the divergent paths of the American people from 1800 to the mid-1800s and the challenges they faced, with emphasis on the Northeast.

Key Terms and People

transcendentalism belief that people should rise above material things in life and that people should depend on themselves rather than outside authority

Ralph Waldo Emerson American writer most famous for his essay "Self-Reliance"

Margaret Fuller American writer who wrote and edited material on transcendentalism

Henry David Thoreau American writer most famous for the transcendental ideas he summarized in his book *Walden*

utopian communities experimental communities that tried to create a perfect society

Nathaniel Hawthorne American writer best known for his novel *The Scarlet Letter*

Edgar Allan Poe American writer best known for his short stories and poetry

Emily Dickinson American poet whose short-style poems were published after her death

Henry Wadsworth Longfellow American poet who wrote popular story-poems such as *The Song of Hiawatha*

Walt Whitman American poet best known for his poem *Leaves of Grass*

Academic Vocabulary

abstract expressing a quality or idea without reference to an actual thing

Section Summary

TRANSCENDENTALISM AND UTOPIAN COMMUNITIES

Transcendentalism was a belief system in which followers thought they could rise above material things in life and that people should depend on themselves rather than outside authority. **Ralph Waldo Emerson, Margaret Fuller**, and **Henry David Thoreau** were among the great American thinkers who were also transcendentalists. Emerson expressed his ideas in the essay "Self-Reliance." Fuller wrote *Women in the Nineteenth Century*, a

Name two transcende[nt]
authors.

_____

_____

Copyrigh[t]
Chapter

book about women's basic rights. In his book *Life in the Woods,* Thoreau summarized many of his transcendentalist beliefs.

Some transcendentalists created communities apart from society. In these **utopian communities**, people hoped to form a perfect society. Some, such as the Shaker communities, were based on religious beliefs. Other groups pursued utopian lifestyles for other transcendental reasons.

> Why might transcendentalists seek to create utopian communities?
> _____
> _____
> _____

THE AMERICAN ROMANTICS

In the early and mid-1800s, many artists were inspired by simple life and nature's beauty. Some joined the Romantic movement that began in Europe with British poets such as Blake, Byron, Keats, and Shelley. For the Romantics, each person brought a unique point of view to the world. These writers used their emotions to guide their works.

It was at this time that American Romantic writers also began to write in an American style. Female writers such as Ann Sophia Stephens wrote popular historical fiction. Another historical novel, *The Scarlet Letter* by **Nathaniel Hawthorne**, described Puritan life in America. Herman Melville wrote *Moby Dick*, considered to be one of America's finest novels. **Edgar Allan Poe** also gained fame for his short stories and poetry.

> Underline all the artists' names that appear in this section. Put an 'X' by the names you have heard before.

Emily Dickinson, **Henry Wadsworth Longfellow**, and John Greenleaf Whittier are just three of the poets of this time whose works have long outlived them. The same is true of **Walt Whitman**, who used his poem *Leaves of Grass* to praise America's individualism and democracy.

> Circle the titles of famous novels written by American Romantics.

CHALLENGE ACTIVITY

Critical Thinking: Make Inferences In this section, certain writers' names appear in bold print, while other names do not. Explain what you think the difference is. Why is it important to notice the difference?
HSS Analysis Skills CR 3

Chapter 13 New Movements in America

MAIN IDEAS

1. The Second Great Awakening sparked interest in religion.
2. Social reformers began to speak out about temperance and prison reform.
3. Improvements in education affected many segments of the population.
4. Northern African American communities became involved in reform efforts.

 HSS 8.6
Students analyze the divergent paths of the American people from 1800 to the mid-1800s and the challenges they faced, with emphasis on the Northeast.

Key Terms and People

Second Great Awakening late 1700s-early 1800s movement of Christian renewal

Charles Grandison Finney minister who challenged some traditional beliefs

Lyman Beecher minister who spoke against both Charles Grandison Finney and alcohol consumption

temperance movement movement to encourage people not to drink alcohol

Dorothea Dix prison reformer

common-school movement movement to have all children, regardless of background, taught in a common place

Horace Mann education reformer

Catharine Beecher founder of all-female academy in Hartford, Connecticut

Thomas Gallaudet education reformer for the hearing impaired

Section Summary

THE SPIRIT OF REVIVAL

During the 1790s, a period of Christian renewal began. It was known as the **Second Great Awakening**. By the 1830s, it had swept through New England, the Appalachians, and the South.

Charles Grandison Finney was one of the leaders of the Second Great Awakening. Some did not agree with Finney's message. However, the Constitution's First Amendment guaranteed Finney's right to speak and be heard. Through the efforts of Finney and other ministers, many Americans joined churches across the country.

> What can you infer from the fact that this period was called the Second Great Awakening?
>
> _____
>
> _____

SOCIAL REFORMS

In the spirit of the Second Great Awakening, people tried to reform many of society's ills. In the **temperance movement**, people aimed at limiting alcohol consumption. **Lyman Beecher** and other ministers spoke about the evils of alcohol.

Another reformer, **Dorothea Dix**, reported on the terrible conditions she found when she visited some Massachusetts prisons. Imprisoned along with adult criminals were the mentally ill and children. Because of efforts by Dix and others, governments built hospitals for the mentally ill and reform schools for young lawbreakers. They also began to try to reform—not just punish—prisoners.

> **How did prisons change as a result of reformers like Dorothea Dix?**
> _____
> _____
> _____

Education in the early 1800s improved with the **common-school movement**. This movement, led by **Horace Mann**, worked to have all students, regardless of background, taught in the same place. Women's education also improved at this time. Several women's schools, including **Catharine Beecher's** all-female academy in Connecticut, opened. Teaching people with disabilities improved, too. For example, **Thomas Gallaudet** bettered the education of the hearing impaired.

> **What was the common-school movement?**
> _____
> _____

AFRICAN AMERICAN COMMUNITIES AND SCHOOLS

In this period, life improved for the nation's free black population. The Free Africans Religious Society, founded by **Richard Allen**, pressed for equality and education. Leaders such as Alexander Crummel helped build African American schools in New York, Philadelphia, and other cities. In 1835 Oberlin College became the first college to admit African Americans. Soon after, in the 1840s, several African American colleges were founded.

> **Circle the names of all the reformers who worked to better America during this time.**

CHALLENGE ACTIVITY

Critical Thinking: Compare and Contrast What did Horace Mann, Catharine Beecher, Thomas Gallaudet, and Richard Allen all have in common? **HSS Analysis Skills HI 1**

Chapter 13 New Movements in America

MAIN IDEAS

1. Americans from a variety of backgrounds actively opposed slavery.
2. Abolitionists organized the Underground Railroad to help enslaved Africans escape.
3. Despite efforts of abolitionists, many Americans remained opposed to ending slavery.

 HSS 8.9
Students analyze the early and steady attempts to abolish slavery and to realize the ideals of the Declaration of Independence.

Key Terms and People

abolition complete end to slavery

William Lloyd Garrison abolitionist who ran the *Liberator* newspaper and also helped found the American Anti-Slavery Society

American Anti-Slavery Society organization that wanted immediate emancipation and racial equality

Angelina and Sarah Grimké southern sisters who spoke in favor of abolition

Frederick Douglass ex-slave who became a pro-abolition speaker

Sojourner Truth ex-slave who spoke for abolition and women's rights

Underground Railroad loosely organized group that helped slaves escape from the South

Harriet Tubman ex-slave who freed more than 300 others using the Underground Railroad

Section Summary

ABOLITION

By the 1830s, many Americans formed a movement to end slavery. They supported **abolition**. These abolitionists worked for emancipation, or freedom from slavery, for all who lived in the United States.

Some abolitionists thought that ex-slaves should get the same rights enjoyed by other Americans. Others, however, hoped to send the freed blacks back to Africa to start new colonies there. In fact, the American Colonization Society successfully founded the African colony of Liberia.

Many abolitionists spread the message of abolition using the power of the pen. **William Lloyd**

> What is the difference between abolition and emancipation?
> _____
> _____
> _____

Garrison, for example, ran the *Liberator* newspaper. He also helped found the **American Anti-Slavery Society**. This group believed in emancipation and racial equality. **Angelina and Sarah Grimké** were two sisters from a southern slave-holding family. They wrote pamphlets and a book to try to convince other white people to join the fight against slavery.

AFRICAN AMERICANS FIGHT AGAINST SLAVERY

When **Frederick Douglass** was a slave, he secretly learned to read and write. After he escaped slavery, he used those skills to support the abolition movement by publishing a newspaper and writing books about his life. Douglass also was a powerful speaker who vividly described slavery's horrors. Many other ex-slaves also were active abolitionists. One example was **Sojourner Truth,** who became famous for her anti-slavery speeches.

> Why do you think Frederick Douglass had to learn to read and write in secret?
>
> _____
> _____
> _____

> How were the intended audiences different for the Grimké sisters and Sojourner Truth?
>
> _____
> _____
> _____

THE UNDERGROUND RAILROAD

The **Underground Railroad** was the name given a loosely knit group of white and black abolitionists who held escaped slaves get North to freedom. One of the most famous "conductors" on this Railroad was an ex-slave named **Harriet Tubman**. She made 19 trips to the north, freeing more than 300 slaves.

> What do you think would happen to someone who was caught helping slaves escape?
>
> _____
> _____

OPPOSITION TO ABOLITION

Many white southerners felt slavery was vital to their economy. They also felt that outsiders should not tell them what to do. Some justified enslaving people by claiming that African Americans needed the structure of slavery to survive.

CHALLENGE ACTIVITY

Critical Thinking: Make Inferences Why do you think Frederick Douglass called his newspaper the *North Star?* **HSS Analysis Skills HI 2**

MAIN IDEAS

1. Influenced by the abolition movement, many women struggled to gain equal rights for themselves.

2. Calls for women's rights met opposition from men and women.

3. The Seneca Falls Convention launched the first organized women's rights movement in the United States.

 HSS 8.9
Students analyze the early and steady attempts to abolish slavery and to realize the ideals of the Declaration of Independence.

Key Terms and People

Elizabeth Cady Stanton supporter of women's rights who helped organize the Seneca Falls Convention

Lucretia Mott women's rights supporter who helped organize the Seneca Falls Convention

Seneca Falls Convention the first organized public meeting about women's rights held in the United States

Declaration of Sentiments the document officially requesting equal rights for women

Lucy Stone spokesperson for the Anti-Slavery Society and the women's rights movement

Susan B. Anthony women's rights supporter who argued for equal pay for equal work, the right of women to enter traditionally male professions, and property rights

Section Summary

THE INFLUENCE OF ABOLITION

In the mid-1800s, some female abolitionists also began to focus on women's rights in America, despite their many critics. For example, the Grimké sisters were criticized for speaking in public. Their critics felt they should stay at home. Sarah Grimké responded by writing a pamphlet in support of women's rights. She also argued for equal educational opportunities, as well as for laws that treated women in an equal manner.

Abolitionist Sojourner Truth also became a women's-rights supporter. The ex-slave never

Why did critics of the Grimké sisters think women should not speak in public?

_____

_____

learned to read or write, but she became a great and
influential speaker.

OPPONENTS OF WOMEN'S RIGHTS

The women's movement had many critics—both
men and women. Some felt a woman should stay
home. Others felt women were not as physically or
mentally strong as men. Therefore, they needed the
protection of first their fathers, then their husbands.
This was why upon marriage, husbands took control of their wives' property.

> **What did the abolition and women's rights movements have in common?**
>
> _____
>
> _____

THE SENECA FALLS CONVENTION

With the support of such leaders as **Elizabeth
Cady Stanton** and **Lucretia Mott**, the **Seneca Falls
Convention** opened July 19, 1848, in Seneca Falls,
New York. It was the first time American women
organized to promote women's rights. It resulted
in the **Declaration of Sentiments**. This document
officially requested equality for women. It brought
18 charges against men, much as the Declaration of
Independence had brought 18 charges against King
George III.

> **What were some of the rights for which women were fighting?**
>
> _____
>
> _____

THE CONTINUING STRUGGLE

After the convention, more women rose to lead the
fight for rights. **Lucy Stone**, for example, was another abolitionist who spoke out for women's rights. So
did **Susan B. Anthony**. Anthony argued that women
should be paid the same as men for the same job,
and that women could do the jobs reserved for men.
Anthony also fought for property rights for women.
Many states changed their property laws because
of her efforts. But some rights, such as the right to
vote, were not won until much later.

> **Why do you think most of the leaders in the women's rights movement were women?**
>
> _____
>
> _____
>
> _____

CHALLENGE ACTIVITY

Critical Thinking: Evaluate Identify the woman you think had the greatest impact on women's rights. Write a sentence or two explaining your
choice. **HSS Analysis Skills CR 3**

Chapter 14 A Divided Nation

HISTORY–SOCIAL SCIENCE STANDARDS
HSS 8.9 Students analyze the early and steady attempts to abolish slavery and to realize the ideals of the Declaration of Independence.
HSS 8.10 Students analyze the multiple causes, key events, and complex consequences of the Civil War.
HSS ANALYSIS SKILL HR 3 Students distinguish relevant from irrelevant information.

CHAPTER SUMMARY

Northern states fight slavery spreading west.

Kansas becomes a bloody battlefield for pro-slavery and anti-slavery Americans.

The nation breaks in two.

COMPREHENSION AND CRITICAL THINKING

Use the answers to the following questions to fill in the graphic organizer above.

1. Interpret Information What earned Kansas the nickname "Bleeding Kansas?"

_____

_____

2. Draw Conclusions List two events involving Abraham Lincoln that led to the breakup of the nation.

_____

3. Make Inferences Why do you think Lincoln was against the spread of slavery? Explain your answer.

_____

_____

_____

Chapter 14 A Divided Nation

MAIN IDEAS

1. The addition of new land in the West renewed disputes over the expansion of slavery.

2. The Compromise of 1850 tried to solve the disputes over slavery.

3. The Fugitive Slave Act caused more controversy.

4. Abolitionists used antislavery literature to promote opposition.

 HSS 8.10
Students analyze the multiple causes, key events, and complex consequences of the Civil War.

Key Terms and People

popular sovereignty when resident voters make decisions concerning slavery in their region

Wilmot Proviso suggested bill that would outlaw slavery in new U.S. territory

sectionalism situation in which people favor the interests of one region over those of the entire country

Free-Soil Party third political party that formed to support abolition

Compromise of 1850 law that maintained America's slave-state/free-state balance

Fugitive Slave Act law that made it a crime to aid runaway slaves

Anthony Burns Virginia slave-fugitive whose attempted rescue from a Boston jail ended in violence

Uncle Tom's Cabin antislavery novel written by Harriet Beecher Stowe

Harriet Beecher Stowe author of the antislavery novel, *Uncle Tom's Cabin*

Section Summary

THE EXPANSION OF SLAVERY

The nation's debate over slavery continued as the country got bigger. Many northerners, for example, supported the **Wilmot Proviso**, which would outlaw slavery in new lands. Many southerners, on the other hand, did not support the bill. Arguments about the proviso showed how **sectionalism** was dividing the country.

> Why were southerners opposed to the Wilmot Proviso?
> _____
> _____
> _____

Some favored the idea of **popular sovereignty**. They thought each region's voters should decide the question of slavery for that region. The debate was so intense that a third political party, the **Free-Soil Party,** formed to support abolition.

THE COMPROMISE OF 1850

The **Compromise of 1850** was presented by Kentucky's Henry Clay. Its purpose was to maintain the delicate balance between slave and free states. It became law because of support by representatives such as Senator Daniel Webster.

> **What made Henry Clay's law a compromise?**
> _____
> _____
> _____

THE FUGITIVE SLAVE ACT

Part of the Compromise of 1850 required passage of the **Fugitive Slave Act**. This act made it a crime to help runaway slaves. Abolitionists especially reacted in anger to the Compromise. Sometimes that anger turned to violence. This was true when abolitionists tried to rescue Virginia fugitive **Anthony Burns** from a Boston jail.

> **How can you tell that Anthony Burns was a slave?**
> _____
> _____
> _____

ANTISLAVERY LITERATURE

Many abolitionists expressed their antislavery feelings in speeches. Others used the written word to influence people on the issue of slavery. One effective author was **Harriet Beecher Stowe**. In 1852 Stowe's antislavery novel, *Uncle Tom's Cabin*, was published. The book showed some of the consequences of slavery. It sold more than 2 million copies and influenced many to support the end of slavery.

> **How did Harriet Beecher Stowe impact the issue of slavery in America?**
> _____
> _____
> _____

CHALLENGE ACTIVITY

Critical Thinking: Write to Identify Write a paragraph about something you read or saw that made you change your mind. It could be a book, a speech, a television show—even a teacher. **HSS Analysis Skills HI 2**

Chapter 14 A Divided Nation

MAIN IDEAS

1. The debate over the expansion of slavery influenced the election of 1852.

2. The Kansas-Nebraska Act allowed voters to allow or prohibit slavery.

3. Pro-slavery and antislavery groups clashed violently in what became known as "Bleeding Kansas."

 HSS 8.10
Students analyze the multiple causes, key events, and complex consequences of the Civil War.

Key Terms and People

Franklin Pierce Democratic candidate who won the presidential election of 1852

Stephen Douglas representative who introduced what would become the Kansas-Nebraska Act

Kansas-Nebraska Act the law that divided the rest of the Louisiana Purchase into two territories—Kansas and Nebraska

Pottawatomie Massacre the murder of five pro-slavery men at Pottawatomie Creek by John Brown and several other abolitionists

Charles Sumner Massachusetts senator who was an outspoken critic of pro-slavery leaders

Preston Brooks South Carolina representative who used a cane to beat Charles Sumner on the Senate floor for his criticisms of pro-slavery leaders

Academic Vocabulary

implications things inferred or deduced

Section Summary

THE ELECTION OF 1852

In the presidential election of 1852, the Democrats nominated **Franklin Pierce**. He was not a well-known politician, however his promise to honor the Compromise of 1850 assured him many southern votes. Pierce ran against Whig candidate Winfield Scott.

Pierce's win over Scott was resounding. When the votes were counted, it was discovered that out of the 31 states, 27 voted for Pierce.

> Why was Franklin Pierce a popular candidate in the South?
>
> _____
>
> _____

THE KANSAS-NEBRASKA ACT

The slavery issue continued to plague the United States. In 1854 Representative **Stephen Douglas** introduced a bill that addressed slavery in the Louisiana Territory. When it was signed into law on May 30, it became known as the **Kansas-Nebraska Act**. It got its name from the two territories into which it divided the rest of Louisiana—Kansas and Nebraska. In each territory, popular sovereignty would determine the answer to the slavery question.

> How did the Kansas-Nebraska Act get its name?
> _____
> _____
> _____

To make sure Kansas voted in favor of slavery, pro-slavery voters left their homes in Missouri to cross the border and vote in Kansas. They won and quickly set up a pro-slavery government. However, those who did not believe in slavery set up another, separate government in Topeka.

> What do you think would be the consequences of one state having two governments?
> _____
> _____

"BLEEDING KANSAS"

In May 1856, pro-slavery jurors charged antislavery leaders with treason. Pro-slavery forces rode to Lawrence to arrest those charged. When they found the suspects had fled, they burned and looted the town.

> What caused the Sack of Lawrence?
> _____
> _____
> _____
> _____

The Sack of Lawrence outraged many abolitionists, including New England abolitionist John Brown. Together with a small group that included four of his sons, Brown was responsible for the **Pottawatomie Massacre**, in which five pro-slavery men were killed. Quickly, Kansas fell into civil war.

Fighting even took place on the Senate floor. South Carolina Representative **Preston Brooks** used his cane to beat Massachusetts Senator **Charles Sumner** into unconsciousness because of Sumner's criticisms of pro-slavery leaders.

> Was Senator Charles Sumner for or against slavery?
> _____
> _____
> _____

CHALLENGE ACTIVITY

Critical Thinking: Write to Explain Write a few sentences to explain how Kansas got the nickname "Bleeding Kansas." **HSS Analysis Skills: HI1**

MAIN IDEAS

1. Political parties in the United States underwent change due to the movement to expand slavery.

2. The *Dred Scott* decision created further division over the issue of slavery.

3. The Lincoln-Douglas debates brought much attention to the conflict over slavery.

 HSS 8.10
Students analyze the multiple causes, key events, and complex consequences of the Civil War.

Key Terms and People

Republican Party political party founded to fight slavery

James Buchanan Democratic candidate and winner of the 1856 presidential election

John C. Frémont Republic candidate for the 1856 presidential election

Dred Scott slave who unsuccessfully sued for his freedom in 1846

Roger B. Taney Chief Justice of the Supreme Court during the *Dred Scott* decision

Abraham Lincoln early leader of the Republican Party

Lincoln-Douglas debates debates between senatorial candidates Abraham Lincoln and Stephen Douglas

Freeport Doctrine Stephen Douglas's belief in popular sovereignty, stated during the Freeport debate

Academic Vocabulary

complex difficult, not simple

Section Summary

NEW DIVISIONS

As the 1850s progressed, Whigs, Democrats, Free-Soilers, and abolitionists united to create the **Republican Party** to fight slavery. Others left their parties to form the Know-Nothing Party. For the 1856 presidential election, the old Democratic Party nominated **James Buchanan**. Buchanan had been out of the country during the Kansas bloodshed, but he defeated Know-Nothing Millard Fillmore and Republican **John C. Frémont**.

> Why would it matter to voters that James Buchanan had been out of the country during "Bleeding Kansas?"
>
> _____
> _____
> _____

THE *DRED SCOTT DECISION*

Dred Scott was a slave. His slaveowner was a doctor who traveled from Missouri, a slave state, to free areas and back again to Missouri. Scott sued for his freedom, since he had lived in free states.

The Supreme Court's decision was against Scott. Chief Justice **Roger B. Taney** wrote that African Americans were not citizens, and only citizens could sue in federal court. Taney also wrote that slaves were considered property, and Scott living in free territory did not make him free. Taney said that Congress could not stop people from taking slaves into federal territory.

Many antislavery voices rose against the decision. This included the voice of an Illinois lawyer named **Abraham Lincoln**.

> Underline the three decisions the Supreme Court made in the *Dred Scott* case.

> Are you surprised to know that at the time of the *Dred Scott* decision, a majority of Supreme Court Justices were from the South? Why or why not?
>
> _____
> _____
> _____

THE LINCOLN-DOUGLAS DEBATES

In 1858 Abraham Lincoln ran for a U.S. Senate seat as the Republican candidate. His opponent was Democrat Stephen Douglas, who was up for re-election. During the campaign, the two men met several times in what became known as the **Lincoln-Douglas debates**. In the debates, Lincoln was careful not to talk about slavery in the existing slave states. Instead, he claimed the Democrats were trying to spread slavery across the nation.

During the second debate, Lincoln questioned Douglas about popular sovereignty. He wondered whether that belief went against the *Dred Scott* decision. In other words, how could the people ban what the Supreme Court allowed? Douglas restated his belief in popular sovereignty. His response was remembered as the **Freeport Doctrine**.

> Why do you believe Lincoln would not talk about slavery in the existing slave states?
>
> _____
> _____
> _____

> Why did Lincoln question the Democrats' belief in popular sovereignty?
>
> _____
> _____
> _____

CHALLENGE ACTIVITY

Critical Thinking: Write to Summarize Write a paragraph summarizing the impact of the Fifth Amendment on the Supreme Court's ruling in the *Dred Scott* case. **HSS Analysis Skills: HI 1**

Chapter 14 A Divided Nation

MAIN IDEAS

1. John Brown's raid on Harpers Ferry intensified the disagreement between free states and slave states.

2. The outcome of the election of 1860 divided the United States.

3. The dispute over slavery led the South to secede.

 HSS 8.10
Students analyze the multiple causes, key events, and complex consequences of the Civil War.

Key Terms and People

John Brown's raid Brown's attack on the Harpers Ferry arsenal, which began October 16, 1859

John C. Breckinridge pro-slavery candidate nominated by southern Democrats for the 1860 presidential election

Constitutional Union Party new political party that concentrated on constitutional principles

John Bell candidate nominated for the 1860 election by the Constitutional Union Party

secession formal withdrawal from the Union

Confederate States of America the country formed by seceding southern states

Jefferson Davis the Confederacy's first president

John J. Crittenden Kentucky senator who proposed a compromise to try to stop southern secession

Section Summary

THE RAID ON HARPERS FERRY

John Brown was an abolitionist. He decided to use violence to try to stop slavery. He planned to lead an attack on the arsenal at Harpers Ferry, Virginia.

    John Brown's raid began on October 16, 1859. Although he succeeded in taking the arsenal, federal troops overwhelmed him and his small band. They killed some of Brown's followers and captured others, including Brown himself.

> When did John Brown's raid begin?
>
> _____
>
> _____

JUDGING JOHN BROWN

Brown was charged and found guilty. On December 2, 1859, he was hanged for his crimes.

Many northerners agreed with Brown's anti-slavery beliefs, but they did not agree with his violent methods. Southerners worried that Brown's raid was the start of more attacks on the South.

> Why do you think John Brown's raid scared southerners?
> _____
> _____

THE ELECTION OF 1860

The country was torn as the 1860 presidential election approached. The Democrats proposed two candidates—the North's Stephen Douglas and the South's **John C. Breckinridge**. In addition, the new **Constitutional Union Party** nominated **John Bell**. Abraham Lincoln ran on the Republican ticket.

Lincoln won the election, but he did not carry a southern state in his win. This angered southerners, who worried that they had lost their political power.

> Underline the names of the presidential candidates who ran for election in 1860.

THE SOUTH SECEDES

Southern states responded to Lincoln's election with **secession**. These states joined together in a new country—the **Confederate States of America**. They elected Mississippian **Jefferson Davis** as their first president. In this country, slavery was legal.

Lincoln argued that southern states could not secede. It seemed that even compromises, like one proposed by Kentucky Senator **John J. Crittenden**, would not mend this tear in the national fabric.

> Why did southern states secede from the United States of America?
> _____
> _____
> _____
> _____

THE NORTH RESPONDS

President-elect Lincoln declared there could be no compromise where slavery was concerned. He also announced that the federal property in southern lands remained part of the United States.

CHALLENGE ACTIVITY

Critical Thinking: Write to Analyze Write a paragraph explaining why the Democrats ran two candidates in the 1860 presidential election and the effect that had on the South's secession. **HSS Analysis Skills CS 1**

Chapter 15 The Civil War

HISTORY–SOCIAL SCIENCE STANDARDS
HSS 8.10 Students analyze the multiple causes, key events, and complex consequences of the Civil War.

CHAPTER SUMMARY

Soldiers, North and South

Union Soldier   **Both Soldiers**   **Confederate Soldier**

official uniform was blue   mostly farmers   from the southern, pro-slavery states

COMPREHENSION AND CRITICAL THINKING

Use information from the graphic organizer to answer the following questions.

1. **Compare and Contrast** Copy and complete this Venn diagram by adding one additional phrase or sentence to each area of the diagram.

   _____

2. **Make Generalizations** After completing the diagram, what generalization can you make about northern states and the slavery issue?

   _____

   _____

3. **Make Connections** What color uniforms would soldiers from the border states wear?

   _____

   _____

Chapter 15 The Civil War

> **MAIN IDEAS**
> 1. Following the outbreak of war at Fort Sumter, Americans chose sides.
> 2. The Union and the Confederacy prepared for war.

 HSS 8.10
Students analyze the multiple causes, key events, and complex consequences of the Civil War.

Key Terms and People

Fort Sumter federal post in Charleston, South Carolina, that surrendered to the Confederacy

border states the four slave states—Delaware, Kentucky, Maryland, and Missouri—that bordered the North

Winfield Scott Union general with a two-part strategy for defeating the Confederacy

cotton diplomacy Confederate plan to enlist England's aid in return for continued cotton shipments

Section Summary

LINCOLN FACES A CRISIS

By the time Abraham Lincoln took office in 1861, seven southern states had already left the Union. He promised he would not attack the southern states. However, he also promised to preserve the Union.

How did the South react to Lincoln's election to the presidency?

_____

_____

_____

Confederate officials already were taking control of federal mints, storehouses, and forts. Fighting finally broke out at **Fort Sumter**, a federal fort in the Confederate state of South Carolina. Federal troops refused to surrender to the Confederacy, but within two days, Fort Sumter fell. Lincoln called for 75,000 troops to put down the South's rebellion.

How did Lincoln respond to the surrender of Fort Sumter?

_____

_____

_____

CHOOSING SIDES

After Lincoln called for troops, all the states had to choose a side. Four more slave states joined the Confederacy. Four **border states**—slave states that bordered the North—decided to stay in the Union.

In addition, western Virginia broke off from Confederate Virginia to stay loyal to the Union.

Union General **Winfield Scott** had a two-part strategy to conquer the South. He would destroy its economy with a naval blockade. He also would gain control of the Mississippi River to help divide the Confederacy.

The Confederacy had its own plan of attack. Part of that plan involved **cotton diplomacy**—the hope that Britain would support the Confederacy because it needed Confederate cotton. Instead, Britain found new cotton suppliers in other lands, such as India and Egypt.

Why didn't cotton diplomacy work?

_____
_____
_____
_____

FROM CIVILIAN TO SOLDIER

Neither side was prepared when it came to war. However, many citizens—northern and southern— were eager to help. Thousands upon thousands of young men answered the call to arms and volunteered to serve in both armies.

Civilians, too, volunteered to help. They raised money to aid soldiers and their families. They staffed and supplied emergency hospitals. In the North alone, about 3,000 women served as army nurses.

In what ways did civilians help the war effort?

_____
_____
_____
_____

Once the thousands of farmers, teachers, merchants, laborers, and others joined the armies, they had to be trained to become soldiers. For long days they drilled and practiced with their guns and bayonets. Soon, they were ready to fight.

CHALLENGE ACTIVITY

Critical Thinking: Write to Influence Write an advertisement encouraging people to support the soldiers by coming to a bandage-wrapping event. **HSS Analysis Skills HI 1**

MAIN IDEAS

1. Union and Confederate forces fought for control of the war in Virginia.
2. The Battle of Antietam gave the North a slight advantage.
3. The Confederacy attempted to break the Union naval blockade.

 HSS 8.10
Students analyze the multiple causes, key events, and complex consequences of the Civil War.

Key Terms and People

Thomas "Stonewall" Jackson inspirational Confederate general who helped fight back Union troops at the First Battle of Bull Run

First Battle of Bull Run 1861 battle near Manassas Junction, Virginia

George B. McClellan general sent by President Lincoln to capture Richmond

Robert E. Lee became head of the Confederate army in June 1862

Seven Days' Battles series of five battles that ended with Lee forcing McClellan and his troops to retreat from the area

Second Battle of Bull Run surprise Confederate attack that helped push the Union forces out of Virginia

Battle of Antietam Maryland battle that resulted in Lee's forces retreating to Virginia

ironclads ships that were heavily armored with iron

Academic Vocabulary

innovation a new idea or way of doing something

Section Summary

TWO ARMIES MEET

It was July 1861. The Union and Confederate armies clashed about 30 miles outside of Washington, D.C., near Manassas Junction, Virginia, along Bull Run Creek. At first the Union soldiers, under General Irvin McDowell, pushed back the left side of the Confederate line. Southern troops, inspired by General **Thomas "Stonewall" Jackson**, fought back. With reinforcements arriving, the Confederate troops drove the Union army back. The **First Battle of Bull Run**, as this conflict would be called, showed that there would be no easy victory.

> From where did the First Battle of Bull Run get its name?
>
> _____
> _____
> _____

MORE BATTLES IN VIRGINIA

The Battle of Bull Run disappointed President Lincoln. He had hoped the Union troops could take the Confederate capital of Richmond, Virginia.

Lincoln tried again, this time sending General **George B. McClellan**. By April 1862, McClellan and 100,000 soldiers were assembled outside Richmond. The Union took Yorktown and forced the southern army to retreat.

Then in June, General **Robert E. Lee** took command of the Confederate forces. On June 26, the two armies met in the first of five clashes that would take place over the next week. The **Seven Days' Battles**, as they were called, resulted in Lee pushing McClellan away from Richmond. Later, at the **Second Battle of Bull Run**, Lee's troops surprised the enemy, pushing them out of Virginia.

> Why did President Lincoln keep sending troops into Virginia?
> _____
> _____
> _____
> _____

> Underline all the Civil War battles identified in this reading.

THE BATTLE OF ANTIETAM

A battle plan accidentally left behind by southern troops led to the next major battle of the Civil War. The Union discovered the Confederates were going to attack Harpers Ferry. McClellan sent his troops to stop them.

The **Battle of Antietam** took place on September 17, 1862. It ended Lee's attempt to take the war to the North. It also ended McClellan's command, for he failed to follow Lincoln's orders to pursue the beaten Confederates.

> Why did President Lincoln take away General McClellan's command?
> _____
> _____
> _____
> _____

THE BLOCKADE

Despite the distance it had to monitor, the Union blockade of Southern ports was very effective. Even though both sides developed **ironclads** at the same time, the sea remained in Union control.

CHALLENGE ACTIVITY

Critical Thinking: Write to Explain How was a mistake responsible for the Battle of Antietam? **HSS Analysis Skills HI 4**

Chapter 15 The Civil War

Section 3

MAIN IDEAS

1. Union strategy in the West centered on control of the Mississippi River.

2. Confederate and Union troops struggled for dominance in the Far West.

 HSS 8.10
Students analyze the multiple causes, key events, and complex consequences of the Civil War.

Key Terms and People

Ulysses S. Grant Union general whose troops won several important battles on southern soil

Battle of Shiloh a two-day battle won by Union troops, even though Confederates had caught them by surprise

David Farragut daring naval leader who helped the Union take control of New Orleans

Siege of Vicksburg six-week blockade of Vicksburg that starved the city into surrender

Section Summary

WESTERN STRATEGY

In February 1862, General **Ulysses S. Grant** led a Union army into Tennessee. He was headed toward the Mississippi River, where Union control would separate the eastern Confederacy from its western, food-supplying states. On the way, Grant and his forces took both Fort Henry and Fort Donelson. Then he moved south to Pittsburg Landing on the west bank of the Tennessee River.

Just north of the Mississippi border, near a church called Shiloh, Grant halted his troops to wait for more soldiers to arrive. Although Grant was aware of Confederate troops in the area, he was caught by surprise when they attacked on April 6.

On the first day of the **Battle of Shiloh**, the Confederates had the advantage. In the night though, Union reinforcements arrived. By the next night, the Confederates were retreating. This win helped the Union control part of the Mississippi River Valley.

> Why did the Union consider control of the Mississippi River critical?
>
> _____
> _____
> _____
> _____

> How do you know that Fort Henry and Fort Donelson were Confederate forts?
>
> _____
> _____
> _____
> _____

FIGHTING FOR THE MISSISSIPPI RIVER

To control the Mississippi River, the Union had to capture New Orleans, the south's largest city and the valuable port near the mouth of the Mississippi River. However, two forts kept ships from approaching New Orleans from the south.

Union naval leader **David Farragut** solved that problem by racing past the two forts in the darkness before dawn on April 24, 1862. Within days, New Orleans had surrendered. Farragut continued north, taking more and more control of the river, until he reached Vicksburg, Mississippi.

The bluffs above Vicksburg had allowed Confederate General John C. Pemberton to effectively stop any attempt to take the city. So instead of trying to capture Vicksburg, General Grant laid siege to the city. The **Siege of Vicksburg** lasted about six weeks before hunger forced the Confederates to surrender. Now the Mississippi River was under Union control.

How was New Orleans captured?
_____
_____
_____
_____

Why was Vicksburg difficult to capture?
_____
_____
_____
_____

Why did the Siege of Vicksburg succeed when attacks on Vicksburg had failed?
_____
_____

THE FAR WEST

Fighting also broke out in the southwest, as the Confederates tried to take control there. Defeats in New Mexico and Arizona ended dreams of a Confederate southwest. Confederate-Union conflict in Missouri also ended with a Confederate defeat.

CHALLENGE ACTIVITY

Critical Thinking: Write to Analyze Write a paragraph analyzing why the South wanted to control the Southwest. **HSS Analysis Skills HI 1**

Chapter 15 The Civil War

MAIN IDEAS

1. The Emancipation Proclamation freed slaves in Confederate states.

2. African Americans participated in the war in a variety of ways.

3. President Lincoln faced opposition to the war.

4. Life was difficult for soldiers and civilians alike.

 HSS 8.10

Students analyze the multiple causes, key events, and complex consequences of the Civil War.

Key Terms and People

emancipation the act of being freed from slavery

Emancipation Proclamation President Lincoln's announcement freeing Confederate slaves

contrabands escaped slaves

54th Massachusetts Infantry African American troop of soldiers that distinguished itself during the Civil War

Copperheads enemies' nickname for the Peace Democrats

habeas corpus the constitutional protection against unlawful imprisonment

Clara Barton army volunteer whose work became the basis for the American Red Cross

Section Summary

THE EMANCIPATION PROCLAMATION

President Lincoln realized that freeing the slaves, or **emancipation**, would take away the slave system on which the South's economy relied. Lincoln presented the **Emancipation Proclamation**, which freed all Confederate slaves. The exceptions were slaves in the border states, which Lincoln did not want to anger.

> What was the purpose of the Emancipation Proclamation?
>
> _____
> _____
> _____
> _____

AFRICAN AMERICANS AND THE WAR

In July 1862, Congress decided to allow both free African Americans and **contrabands**, escaped slaves, to join the army. The most famous African American unit was the **54th Massachusetts Infantry**, which helped attack South Carolina's Fort Wagner.

African American soldiers received less pay than white soldiers and faced greater danger if captured. Lincoln suggested these soldiers be rewarded by getting the right to vote.

> How were contrabands different from other African Americans who joined the Union army?
> _____
> _____

POLITICAL CHALLENGES TO THE WAR

Not everyone agreed with the war. Some midwesterners, for example, did not think the war was necessary. They called themselves Peace Democrats. But their enemies called them **Copperheads**, after the poisonous snake.

To silence their protests, Lincoln had them put in jail with no evidence and no trial. This meant he ignored their right of **habeas corpus**. This is the constitutional protection against unlawful imprisonment. Despite this and other controversies, Lincoln won his second election in 1864.

> What is habeas corpus?
> _____
> _____
> _____

THE LIVES OF SOLDIERS

For the soldier, both camplife and combat offered dangers. Poor camp conditions led to illness, which killed more men than did battle. Those wounded or captured in battle often met the same fate.

> Which led to more deaths, illness or battles?
> _____
> _____

LIFE ON THE HOME FRONT

Those left behind took over the work of the men who went to war. In addition, many women also provided medical care for the soldiers. For example, volunteer **Clara Barton** formed the organization that would become the American Red Cross.

> How did women help the war effort?
> _____
> _____
> _____

CHALLENGE ACTIVITY

Critical Thinking: Write to Debate You are a lawyer for the Peace Democrats. Write a paragraph explaining why their right of habeas corpus should not be ignored. **HSS Analysis Skills HI 2**

Section 5

MAIN IDEAS
1. The Battle of Gettysburg was an important turning point in the war.
2. During 1864, Union campaigns in the East and South dealt crippling blows to the Confederacy.
3. Union troops forced the South to surrender, ending the Civil War.

 HSS 8.10
Students analyze the multiple causes, key events, and complex consequences of the Civil War.

Key Terms and People

George G. Meade general who led Union troops during the Battle of Gettysburg

Battle of Gettysburg three-day battle at Gettysburg that ended with a Confederate loss

George Pickett general who carried out Lee's orders to charge the Union line

Pickett's Charge disastrous attempt of Pickett's troops to storm Cemetery Ridge

Gettysburg Address speech in which Lincoln dedicated Gettysburg's new cemetery

Wilderness Campaign series of Virginia battles in which General Grant tried in vain to take Richmond

William Tecumseh Sherman Union general who cut a 250-mile path of destruction across Georgia

total war strategy in which troops destroy both civilian and military resources

Appomattox Courthouse the place where General Lee surrendered to General Grant

Academic Vocabulary

execute perform, carry out

Section Summary

THE BATTLE OF GETTYSBURG

In December 1862, Confederate troops under the command of General Robert E. Lee won a battle at Fredericksburg, Virginia. Later, they headed into Union territory. Lee hoped a Confederate win on Union soil would break the Union's spirit.

The northern battle came at Gettysburg, Pennsylvania, where Confederate and Union troops surprised each other. The **Battle of Gettysburg** started July 1, 1863. It lasted three days. On the

> How long did the Battle of Gettysburg last?
>
> _____
>
> _____

first day, Lee's troops pushed back General **George
G. Meade's** Union soldiers. The Union troops had
to dig in on top of two hills outside the town. On
the second day, Confederates tried to take the hill,
called Little Round Top. They failed.

On the third day, Lee ordered General **George Pickett**
to charge his men up Cemetery Ridge. **Pickett's Charge**
was a disaster. More than half of the Confederates were
killed before the charge ended. Lee retreated. Never
again would his troops reach northern land.

President Lincoln helped dedicate the new soldier
cemetery at Gettysburg. On November 19, 1863,
he delivered the **Gettysburg Address**, in which he
restated the purpose of the war.

> Who won the Battle of
> Gettysburg?
> _____
> _____

GRANT'S DRIVE TO RICHMOND

The **Wilderness Campaign** was a series of battles
fought in Virginia around Richmond. Although
Grant lost more men than Lee, he had more to lose.
So he continued forward. Grant was winning the
war, but he failed to capture Richmond.

> Why was the capture of Richmond
> such an important goal for the
> Union?
> _____
> _____
> _____
> _____

SHERMAN STRIKES THE SOUTH

To assure his re-election, Lincoln needed a victory.
He sent General **William Tecumseh Sherman** into
Georgia. Sherman fought his way to Atlanta and
then bombed and burned most of the town. Lincoln
was re-elected in a landslide. Sherman then ordered
his troops to cut a path of destruction through
Georgia, practicing **total war**—that is, destroying
everything—all the way to the sea.

> How did General Sherman help
> President Lincoln get
> re-elected?
> _____
> _____
> _____

THE SOUTH SURRENDERS

On April 9, 1865, at **Appomattox Courthouse**, Lee
officially surrendered to Grant. The war was over.
Rebuilding could begin.

> How long did the Civil War last?
> _____
> _____

CHALLENGE ACTIVITY

Critical Thinking: Make a Timeline Use the dates and events in this reading
to make a time line titled "The Civil War." **HSS Analysis Skills CS 2, HI 2**

Chapter 16 Reconstruction

HISTORY–SOCIAL SCIENCE STANDARDS

HSS 8.10 Students analyze the multiple causes, key events, and complex consequences of the Civil War.

HSS 8.11 Students analyze the character and lasting consequences of Reconstruction.

HSS ANALYSIS SKILLS CS 3 Students distinguish relevant from irrelevant information, essential from incidental information, and verifiable from unverifiable information in historical narratives and stories.

HSS ANALYSIS SKILLS HI 1 Students explain the central issues and problems from the past, placing people and events in a matrix of time and place.

CHAPTER SUMMARY

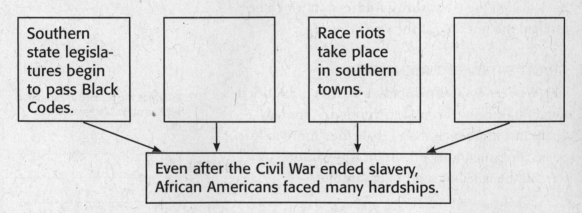

Southern state legislatures begin to pass Black Codes.

Race riots take place in southern towns.

Even after the Civil War ended slavery, African Americans faced many hardships.

COMPREHENSION AND CRITICAL THINKING

Use the answers to the following questions to fill in the graphic organizer above.

1. Summarize Describe four things that made post-war life difficult for African Americans.

_____

_____

2. Make Inferences At which level of government do you think African Americans found the least discrimination during Reconstruction? Explain your answer.

_____

_____

Chapter 16 Reconstruction

> **MAIN IDEAS**
> 1. President Lincoln and Congress differed in their views as Reconstruction began.
> 2. The end of the Civil War meant freedom for African Americans in the South.
> 3. President Johnson's plan began the process of Reconstruction.

 HSS 8.11
Students analyze the character and lasting consequences of Reconstruction.

Key Terms and People

Reconstruction period of reuniting and rebuilding the South following the end of the Civil War

Ten Percent Plan Lincoln's Reconstruction plan called for 10 percent of voters in a state to pledge loyalty to the United States before that state could rejoin the Union

Thirteenth Amendment the amendment that made slavery illegal throughout the United States

Freedmen's Bureau an organization established by Congress to provide relief for all the South's poor people

Andrew Johnson vice president who became president upon Lincoln's death

Academic Vocabulary

procedure a series of steps taken to accomplish a task

Section Summary

PLANNING RECONSTRUCTION

As soon as the Civil War ended, Reconstruction began. **Reconstruction** was the process of reuniting the nation and rebuilding the southern states.

President Lincoln proposed that southerners be offered amnesty, or an official pardon. All southerners had to do was swear an oath of loyalty to the United States and accept a ban on slavery. When 10 percent of the voters in any state took the oath, that state could be accepted back into the Union. This was called the **Ten Percent Plan**.

Some supported the Wade-Davis bill instead. The Wade-Davis bill also called for southerners to ban

> What did the Ten Percent Plan and the Wade-Davis bill have in common?
>
> _____
> _____
> _____
> _____

slavery. However, under this bill, most of the people
of a state would have to take the pledge before the
state could rejoin the Union. Also, only southerners
who swore they did not support the Confederacy
could run for office. Lincoln vetoed it.

FREEDOM FOR AFRICAN AMERICANS

In 1865, the **Thirteenth Amendment** to the
Constitution officially outlawed slavery in the
nation. Former slaves responded to freedom in
many ways. They legalized their marriages, searched
for relatives who had been sold away, took last
names, and moved to new places.

> What part of the Constitution
> granted freedom to all slaves?
> _____
> _____

THE FREEDMEN'S BUREAU

In 1865 Congress created the **Freedmen's Bureau** to
help all the South's poor people. One of its roles was
to build more schools. Some freedpeople also established their own schools. Although some southerners
at times violently resisted the idea of educating African
Americans, freedpeople of all ages attended classes.

> Why would southerners
> oppose the education of African
> Americans?
> _____
> _____
> _____
> _____

PRESIDENT JOHNSON'S RECONSTRUCTION PLAN

On April 14, 1865, President Lincoln was shot while
attending the theater. He died the next morning.
Vice President **Andrew Johnson** became the next
president of the United States.

Johnson's Reconstruction plan included a way to
restructure southern state governments. States that
followed the steps were readmitted to the Union.

Though the southern states did as Johnson
requested, Congress refused to accept them back
into the Union because many of the representatives
had been Confederate leaders. Clearly, there were
still problems to be solved.

> Why did Congress refuse to accept
> the southern states back into the
> Union?
> _____
> _____
> _____
> _____

CHALLENGE ACTIVITY

Critical Thinking: Write to Defend You are a citizen from a southern
state. Write a letter to Congress, defending your right to choose your
state's representatives. **HSS Analysis Skills: CR 5**

Chapter 16 Reconstruction

Section 2

MAIN IDEAS
1. Black Codes led to opposition to President Johnson's plan for Reconstruction.
2. The Fourteenth Amendment ensured citizenship for African Americans.
3. Radical Republicans in Congress took charge of Reconstruction.
4. The Fifteenth Amendment gave African Americans the right to vote.

 HSS 8.11
Students analyze the character and lasting consequences of Reconstruction.

Key Terms and People

Black Codes southern laws greatly limiting the freedom of African Americans

Radical Republicans Republicans who wanted more federal control in Reconstruction

Civil Rights Act of 1866 act giving African Americans the same legal rights as whites

Fourteenth Amendment amendment guaranteeing citizens equal protection of the laws

Reconstruction Acts laws passed to protect African American rights

impeachment process of bringing charges of wrongdoing against a public official

Fifteenth Amendment amendment guaranteeing suffrage to African American men

Academic Vocabulary

principle basic belief, rule, or law

Section Summary

THE BLACK CODES

Almost as soon as the southern states created new legislatures, those legislatures went to work passing **Black Codes**. The Black Codes were laws that greatly limited the freedom of African Americans. In fact, the codes created working conditions that resembled slavery for African Americans. Protests against the codes were ineffective.

> Why do you think the Black Codes were passed?
> _____
> _____
> _____
> _____

THE RADICAL REPUBLICANS

The Black Codes angered a group of Republicans known as the **Radical Republicans**. They wanted more federal

Section 2, *continued*

control over Reconstruction to make sure southern leaders did not remain loyal to the Confederacy's ideas. Radical Republican leader Thaddeus Stevens pushed for racial equality. The Radicals branded Johnson's Reconstruction plan a failure.

> Why do you think the Radical Republicans judged President Johnson's Reconstruction plan a failure?
> _____
> _____
> _____
> _____

THE FOURTEENTH AMENDMENT

In 1866 Congress proposed the Freedmen's Bureau bill to give more power to the Freedmen's Bureau. President Johnson vetoed it. He did not believe African Americans needed special assistance.

Then Congress passed the **Civil Rights Act of 1866**. It guaranteed African Americans the same rights as whites. Johnson vetoed this, too. Congress overrode the veto. It also proposed the **Fourteenth Amendment** to secure these protections.

> Why do you think President Johnson vetoed the Civil Rights Act of 1866?
> _____
> _____
> _____
> _____

CONGRESS TAKES CHARGE

After the 1866 elections, Republicans held a two-thirds majority in both the House and Senate. As a result, Congress passed several **Reconstruction Acts**. It also passed a law limiting the president's powers to remove cabinet members without Senate approval. When President Johnson broke that law by firing his secretary of war, Congress responded by impeaching the president. The **impeachment** fell short by one vote, and Johnson remained president, though he had little political power.

THE FIFTEENTH AMENDMENT

Republicans wanted African Americans to support their Reconstruction plan. Republicans in Congress proposed the **Fifteenth Amendment**, which guaranteed African American men the right to vote. This amendment went into effect in 1870.

> After the Fifteenth Amendment was ratified, which Americans still could not vote?
> _____
> _____

CHALLENGE ACTIVITY

Critical Thinking: Research to Discover In the library or on the Internet, read the Fourteenth and Fifteenth Amendments. Write a sentence paraphrasing each amendment. **HSS Analysis Skills CR 4**

Chapter 16 Reconstruction

MAIN IDEAS

1. Reconstruction governments helped reform the South.

2. The Ku Klux Klan was organized as African Americans moved into positions of power.

3. As Reconstruction ended, the rights of African Americans were restricted.

4. Southern business leaders relied on industry to rebuild the South.

 HSS 8.11

Students analyze the character and lasting consequences of Reconstruction.

Key Terms and People

Hiram Revels the first African American senator

Ku Klux Klan secret society that used violence to oppress African Americans

Compromise of 1877 agreement in which Democrats accepted Hayes's election to the presidency in exchange for removing federal troops from the South

poll tax a special tax people had to pay before they could vote

segregation the forced separation of whites and African Americans in public places

Jim Crow laws laws that enforced segregation

Plessy v. Ferguson Supreme Court ruling that upheld segregation

sharecropping system in which farm laborers kept some of the crop

Section Summary

RECONSTRUCTION GOVERNMENTS

After the Civil War, some northern Republicans moved to the South. They were not trusted by southerners, who thought they came to profit from Reconstruction.

African Americans used their new right to vote to elect more than 600 African Americans, including the first black senator, **Hiram Revels**. Together they worked to rebuild the war-damaged South.

> Why do you think African Americans were elected if overall they were in the minority?
>
> _____
> _____
> _____
> _____

OPPOSITION TO RECONSTRUCTION

Many southerners opposed Reconstruction. In 1866 a group of them created the secret and violent **Ku Klux Klan**. Its targets were African Americans,

> Circle the groups of Americans that were targeted by the Ku Klux Klan.

Section 3, *continued*

Republicans, and public officials. The Klan spread throughout the South until the federal government stepped in and passed laws that made Klan activities illegal. Violence, however, continued.

THE ELECTION OF 1876

The General Amnesty Act of 1872 allowed most former Confederates to serve in public office. Soon many Democratic ex-Confederates were elected. Republicans also lost power because of Grant's problem-plagued presidency and the Panic of 1873. In 1876 the Hayes-Tilden presidential race was so close it took the **Compromise of 1877** to make sure Democats would accept Hayes's election.

> Why did southern Republicans lose power during the 1870s?
> _____
> _____
> _____
> _____

Southern Democrats, called Redeemers, worked to limit African American rights. The tools they used included **poll taxes**, legal **segregation**, and **Jim Crow laws**. They even got help from the Supreme Court, which ruled in ***Plessy* v. *Ferguson*** that segregation was legal.

FARMING AND THE "NEW SOUTH" MOVEMENT

Most African Americans could not afford to buy land. So many began **sharecropping**, or sharing the crop with landowners. Often, only the landowner profited, and sharecroppers lived in debt.

The South's economy depended on cotton profits, which went up and down. In the "New South" movement, southern leaders turned toward industry to strengthen the economy. Many new textile mills and factories were built. With cheap labor, industry thrived to help southern economies grow stronger.

> How was the economy of the "Old South" different from the economy of the "New South?"
> _____
> _____
> _____

CHALLENGE ACTIVITY

Critical Thinking: Write to Put in Sequence Write a paragraph explaining how the General Amnesty Act eventually led to the Compromise of 1877. **HSS Analysis Skills HI 2**

Chapter 17 Americans Move West

HISTORY–SOCIAL SCIENCE STANDARDS

HSS 8.8 Students analyze the divergent paths of the American people in the West from 1800 to the mid-1800s and the challenges they faced.

HSS 8.12 Students analyze the transformation of the American economy and the changing social and political conditions in the United States in response to the Industrial Revolution.

ANALYSIS SKILLS

CS 3 Students use a variety of maps and documents to identify physical and cultural features of neighborhoods, cities, states, and countries and to explain the historical migration of people, expansion and disintegration of empires, and the growth of economic systems.

CHAPTER SUMMARY

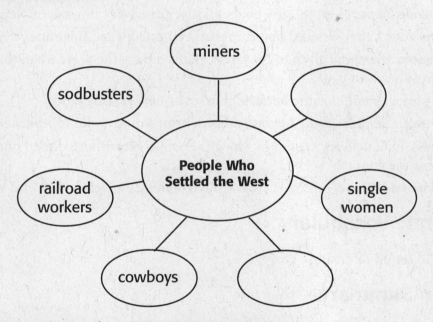

COMPREHENSION AND CRITICAL THINKING

Use the answers to the following questions to fill in the graphic organizer above.

1. **Identify** Look at the graphic organizer above. What other groups came to settle the West?

_____

_____

2. **Make Judgments** Do you think there should be more than seven circles in the settlement web? Why or why not?

_____

_____

MAIN IDEAS

1. A mining boom brought growth to the West.

2. The demand for cattle created a short-lived Cattle Kingdom on the Great Plains.

3. East and West were connected by the transcontinental railroad.

 HSS 8.8

Students analyze the divergent paths of the American people in the West from 1800 to the mid-1800s and the challenges they faced.

Key Terms and People

frontier an area that is undeveloped

Comstock Lode name given to 1859 gold and silver discovery in western Nevada

boomtowns towns that serviced new mines and emptied when the mines closed

Cattle Kingdom the name given to the Great Plains during the years when they supported great herds of cattle

cattle drive movement of cattle herds to market or new grazing lands

Chisholm Trail a popular route for cattle drives from San Antonio to Abilene, Kansas

Pony Express mail delivery system in which messengers transported mail on horseback across the country

transcontinental railroad railroad across United States, connecting East and West

Academic Vocabulary

establish to set up or create

Section Summary

THE GEOGRAPHY AND POPULATION OF THE WEST

After the Civil War, many Americans moved west. Most passed the dry, Indian-populated Great Plains and went on to California or Oregon. By 1850 the **frontier**, or undeveloped region, had reached the Pacific.

New farming techniques and a lessened Indian threat attracted people to the Plains. Newcomers relied on railroads to get their goods to market.

Why did many early pioneers decide not to settle in the Great Plains?

_____

_____

_____

_____

THE MINING BOOM

In 1859 miners found gold and silver in western Nevada. News of the **Comstock Lode** (named after

miner Henry Comstock) brought thousands to
Nevada. Most miners eventually were bought out by
big businesses, which could afford the equipment
needed to remove precious metals from quartz.
Mining jobs were dangerous and low paying. Many
immigrants came to work in the mines. As mines
opened, **boomtowns** grew.

> How did the discovery of precious
> metals affect Nevada's population?
> _____
> _____

THE CATTLE KINGDOM

Demand for beef in the East led to a booming
cattle industry in Texas. Soon the cattle industry
spread onto the Great Plains, creating a huge **Cattle
Kingdom** with giant herds grazing the open range.

> Circle the sentence that helps you
> to infer that cattle need large par-
> cels of land for grazing.

Cowboys, using the techniques of Mexico's
vaqueros, cared for the cattle. They were responsible for the long **cattle drives** that got the cattle to
Abilene, Kansas. There, the new railroad could ship
the cattle to markets far away. The **Chisholm Trail**
was one of the routes they used for this journey.

THE TRANSCONTINENTAL RAILROAD

In 1860 the **Pony Express** improved communication between the East and the West. But it was
replaced in 1861 by the telegraph, which could send
messages much faster.

> Based on what you have read,
> how would you define the word
> *transcontinental*?
> _____

By 1862 Congress started passing laws that supported the building of a **transcontinental railroad**.
By the end of 1863, the laying of tracks had begun,
with one line working from the East and one from
the West. In 1869 the two lines met at Promontory,
Utah. Soon, railroads crisscrossed the country.

> How many years did it take to
> build the transcontinental railroad?
> _____

Railroads provided better transportation for people and goods all over America. By 1890 railroads
were one of the nation's biggest industries.

CHALLENGE ACTIVITY

Critical Thinking: Evaluate Research the Pony Express to find out why
it operated for only 18 months. Write an essay describing what you
learned and giving your opinion about whether the Pony Express was a
success. **HSS Analysis Skills CS 1**

Chapter 17 Americans Move West

MAIN IDEAS

1. As settlers moved to the Great Plains, they encountered the Plains Indians.

2. The United States Army and Native Americans fought in the northern plains, Southwest, and Far West.

3. Despite efforts to reform United States policy toward Native Americans, conflict continued.

 HSS 8.12
Students analyze the transformation of the American economy and the changing social and political conditions in the United States in response to the Industrial Revolution.

Key Terms and People

Treaty of Fort Laramie first major agreement signed with northern Plains nations

reservations areas of federal land set aside for Native Americans

Crazy Horse Sioux leader who violently protested reservations

Treaty of Medicine Lodge treaty in which Plains Indians agreed to live on reservations

buffalo soldiers name given by Indians to U.S. troops that included African American cavalry

George Armstrong Custer army commander who lost to the Sioux at Little Bighorn

Sitting Bull Sioux leader who defeated Custer at Little Bighorn

Battle of the Little Bighorn last great victory for Sioux, in which they defeated Custer

Massacre at Wounded Knee battle in which U.S. troops killed about 150 Sioux

Long Walk a 300-mile forced march of Navajo captives to a New Mexico reservation

Geronimo Chiricahua Apache who battled the U.S. Army to the end

Ghost Dance religious movement predicting a coming paradise for Native Americans

Sarah Winnemucca Paiute who spoke against the government's treatment of Indians

Dawes General Allotment Act act that took back almost 70 percent of reservation land

Section Summary

THE SOUTHERN PLAINS INDIANS

The U.S. government signed the **Treaty of Fort Laramie** to keep peace with the Plains Indians. Later treaties created **reservations** for Indians to live on. Many Indians refused to move to the reservations, and violence followed. In 1866 **Crazy Horse** and his Sioux warriors killed about 81 soldiers. But by the 1867 **Treaty of Medicine Lodge**, most Indians agreed to go

> What was the purpose of early treaties between the U.S. government and Native Americans?
>
> _____
> _____
> _____

to reservations. To force any Native Americans who resisted being moved, the U.S. government sent in **buffalo soldiers**.

In 1874 gold was discovered in the Black Hills of the Dakotas. The government insisted that the Sioux sell their Black Hills reservation. **Sitting Bull** and other Lakota Sioux refused. In response, on June 25, 1876, Lieutenant Colonel **George Armstrong Custer** ordered his soldiers to attack a Sioux camp at the Little Bighorn River. The Sioux won the **Battle of the Little Bighorn**, the worst defeat for the U.S. Army in the West. Fourteen years later, in the last battle against the Plains Indians, 150 Indians were killed in the **Massacre at Wounded Knee**.

> Why do you think that many Native Americans fought the move to reservations?
>
> _____
>
> _____
>
> _____

> Why did the government want the Sioux to sell their reservation?
>
> _____
>
> _____

INDIANS IN THE SOUTHWEST AND FAR WEST

Other Indian groups also fought moving to reservations. U.S. troops forced groups such as the Southwest's Navajo and Oregon's Nez Percé onto reservations far from their homelands. In the **Long Walk** in 1864, Navajo captives were forced to march across 300 miles of desert to a reservation in New Mexico. Many died along the way. In 1886, the Apache leader **Geronimo** and his warrior band surrendered, which ended Apache armed resistance.

PROTEST AND POLICY

In the 1870s, the **Ghost Dance** was a religious movement started by a Paiute Indian named Wovoka. It predicted paradise for Indians. But officials feared this religion would lead to rebellion. Another Paiute, **Sarah Winnemucca**, lectured on problems with the reservation system. In 1887 Congress passed the **Dawes General Allotment Act**, which gave Indians citizenship but took back two thirds of the land originally set aside for them.

> Why would officials worry about the spread of the Ghost Dance beliefs?
>
> _____
>
> _____

CHALLENGE ACTIVITY

Critical Thinking: Write to Evaluate In a paragraph, describe how you think the public reacted to Custer's defeat. **HSS Analysis Skills HI 2**

MAIN IDEAS

1. Many Americans started new lives on the Great Plains.

2. Economic challenges led to the creation of farmers' political groups.

3. By the 1890s, the western frontier had come to an end.

 HSS 8.8

Students analyze the divergent paths of the American people in the West from 1800 to the mid-1800s and the challenges they faced.

Key Terms and People

Homestead Act an 1862 act that gave government-owned land to farmers

Morrill Act an 1862 act that gave government-owned land to the states, with the understanding that the states would sell the land to finance colleges

Exodusters African Americans who left the South for Kansas in 1879

sodbusters nickname given Great Plains farmers because of the toughness of the soil

dry farming farming using hardy crops that need less water than others

Annie Bidwell a Great Plains pioneer-reformer

National Grange a social and educational organization for farmers

deflation a decrease in the money supply and overall lower prices

William Jennings Bryan Democratic candidate for president in the 1888 election

Populist Party new national party formed by the country's farmers

Academic Vocabulary

facilitate to bring about

Section Summary

NEW LANDS FOR SETTLEMENT

Laws passed in 1862 helped open the West to settlers. The **Homestead Act** gave land owned by the federal government to small farmers. The **Morrill Act** granted federal land to the states, which could then sell that land and use the money to build colleges.

Many groups were attracted to the Great Plains, including single women, who could get land through the Homestead Act, and up to 40,000 African Americans. Some called the African Americans **Exodusters** because of their exodus from the South.

> Circle the two congressional acts in 1862 that helped settle the West.

Sodbusters earned their nickname because breaking up the sod of the Great Plains was hard work. **Dry farming** helped the sodbusters succeed, as it meant they chose hardier crops to grow in the dry plains. Cyrus McCormick also helped farmers with his new mechanical farming equipment.

Some pioneer women, such as author Laura Ingalls Wilder and suffragist **Annie Bidwell**, became famous. Most, though, lived a lonely life on the remote farms of the plains.

> How did sodbusters get their nickname?
> _____
> _____

FARMERS FACE CHALLENGES

Machines made farming more productive. This productivity, however, led to overproduction, which in turn led to lower crop prices. To make matters worse, during this time the U.S. economy was affected by **deflation**, in which the money supply decreased and prices dropped. Farmers were thus making less money growing crops. Many lost their farms.

To help each other, farmers formed the **National Grange** (*grange* is an old word for "granary"). The Grange worked to better farmers' lives. It worked to get the railroads regulated. It also backed political candidates such as **William Jennings Bryan**.

> Why could banding together help farmers?
> _____
> _____
> _____

THE POPULIST PARTY

In 1892 groups called Farmers' Alliances formed a new pro-farmer political party. They called it the **Populist Party**. In the 1896 presidential election, it supported the Democratic candidate, Bryan. Bryan lost and the Populist Party dissolved.

CLOSING THE FRONTIER

In 1889, the government opened Oklahoma to homesteaders. These pioneers quickly claimed more than 11 million acres of land. The frontier existed no more.

> What ended America's frontier?
> _____
> _____

CHALLENGE ACTIVITY

Critical Thinking: Write to Connect Explain why overproduction on farms could lead to lower prices for crops. **HSS Analysis Skills HI 2**

Chapter 18 An Industrial Nation

HISTORY–SOCIAL SCIENCE STANDARDS
HSS 8.12 Students analyze the transformation of the American economy and the changing social and political conditions in the United States in response to the Industrial Revolution.
HSS Analysis Skill HI 6 Students interpret basic indicators of economic performance and conduct cost-benefit analyses of economic and political issues.

CHAPTER SUMMARY

CAUSE		EFFECT
The Bessemer Process makes steel quicker and cheaper to produce	⟶	Demand for steel rises
	⟶	Transportation improves
Carnegie, Rockefeller, and other business giants form trusts to monopolize industries	⟶	
Workers work long hours for very little money	⟶	Workers form labor unions

COMPREHENSION AND CRITICAL THINKING

Use the answers to the following questions to fill in the graphic organizer above.

1. Cause and Effect What innovations changed American life in the 1800s?

_____

_____

2. Make Inferences Did the national government approve or disapprove of trusts? How can you tell?

_____

_____

Chapter 18 An Industrial Nation

MAIN IDEAS

1. Breakthroughs in steel processing led to a boom in railroad construction.

2. Advances in the use of oil and electricity improved communications and transportation.

 HSS 8.12

Students analyze the transformation of the American economy and the changing social and political conditions in the United States in response to the Industrial Revolution.

Key Terms and People

Second Industrial Revolution a period of rapid growth in U.S. manufacturing in the late 1800s

Bessemer process Henry Bessemer's invention that made steel production faster and cheaper

Thomas Edison inventor who created the electric lightbulb

patents exclusive rights to make or sell inventions

Alexander Graham Bell inventor of the telephone

Henry Ford introduced the Model T in 1908 and was the first to implement the moving assembly line in manufacturing

Wilbur and Orville Wright brothers who made the first piloted flight in a gas-powered airplane

Academic Vocabulary

implement to put in place

Section Summary

STEEL AND RAILROADS

America's **Second Industrial Revolution** started in the late 1800s. The new **Bessemer process** reduced the amount of time it took to make steel. The price of steel dropped because of this innovation. This made the steel industry an important part of the revolution.

Cheaper, more available steel led to more railroad building. Other changes made train travel far safer and smoother for passengers. Trains strengthened the economy, because they moved goods and people to their destinations quickly and inexpensively.

> **What effect did inexpensive, readily available steel have on the railroad industry?**
>
> _____
> _____
> _____
> _____

USE OF OIL AND ELECTRICITY

In the 1850s scientists figured out how to turn crude oil into kerosene. Kerosene was used for both heat and light. As a result, the demand for oil exploded. In 1859 Edwin L. Drake's Pennsylvania oil well was producing 20 barrels of oil a day. Oil quickly became big business in Pennsylvania, Ohio, and West Virginia.

> What made the demand for oil rise in the 1850s?
> _____
> _____
> _____

RUSH OF INVENTIONS

Thomas Edison was an inventor. In fact, Edison was so productive that he eventually held more than 1,000 **patents** to his inventions.

> How did Edison and Westinghouse help spread the use of electricity?
> _____
> _____
> _____

Edison experimented with electricity. He realized it offered hope as a source of light and power. In 1879 Edison and his assistants created the electric lightbulb. To create a market for his lightbulb, Edison built a power plant to supply industries with electricity. In the late 1880s, George Westinghouse built a power plant that could send electricity far distances to faraway markets. With the help of Edison and Westinghouse, the use of electricity in homes and businesses boomed.

Technological advances changed the way people communicated. First, telegraphs made long-distance communication possible. Then in 1876, inventor **Alexander Graham Bell** unveiled the telephone. By 1900 some 1.5 million telephones were in use.

The late nineteenth century also saw changes in transportation. The invention of a gasoline-powered engine made automobiles possible. In 1908, **Henry Ford** introduced the Model T. The gas-powered engine also allowed **Wilbur and Orville Wright** to invent the airplane.

CHALLENGE ACTIVITY

Critical Thinking: Write to Make Judgments Review all of the inventions about which you just read. In your opinion, which was the most life-changing? Why? **HSS Analysis Skills HI 3**

Chapter 18 An Industrial Nation

MAIN IDEAS

1. The rise of corporations and powerful business leaders led to the dominance of big businesses in the United States.

2. People and the government began to question the methods of big business.

 HSS 8.12

Students analyze the transformation of the American economy and the changing social and political conditions in the United States in response to the Industrial Revolution.

Key Terms and People

corporations businesses owned by stockholders

Andrew Carnegie business leader who concentrated his efforts on steel production

vertical integration owning the businesses involved in each step of manufacturing

John D. Rockefeller business leader who concentrated on oil refining

horizontal integration owning all of the businesses in a certain field

trust a legal arrangement grouping together a number of companies under a single board of directors

Leland Stanford business leader who concentrated on mining equipment and railroads

social Darwinism belief that Charles Darwin's theory of natural selection and "survival of the fittest" holds true for humans

monopoly total ownership of a product or service

Sherman Antitrust Act law that made it illegal to monopolize a business

Academic Vocabulary

acquire to get

Section Summary

THE GROWTH OF BIG BUSINESS

In the late 1800s entrepreneurs began to form **corporations**. A corporation is owned by people who buy shares of its stock. Stockholders share the corporation's profits; but if the corporation fails, stockholders lose the money that they invested. Entrepreneurs could thus spread the risk of loss.

> Why did entrepreneurs form corporations in the late 1800s?
>
> _____
> _____
> _____

BUSINESS LEADERS

One successful entrepreneur of the late 1800s was **Andrew Carnegie**. He made money in several industries, but he focused on steel. Carnegie bought all of the businesses involved in making steel. This process is called **vertical integration**.

John D. Rockefeller made his fortune in oil. Like Carnegie, he used vertical integration. He also used **horizontal integration** and bought many of his competitors. He grouped his companies into a **trust** in an effort to control oil production and prices.

Leland Stanford was another successful business leader who sold mining equipment to miners. He also founded the California Central Pacific railroad.

> **How did entrepreneurs benefit from vertical integration and horizontal integration?**
> _____
> _____
> _____

SOCIAL DARWINISM

In the late 1800s, many people believed in **social Darwinism**. Charles Darwin proposed that in nature, the law was "survival of the fittest." Social Darwinists believed the same was true of humans—those who got rich were the fittest.

Other wealthy business leaders claimed that the rich had a duty to help the poor. As a result, some leaders gave millions of dollars to charities.

> **What is social Darwinism?**
> _____
> _____
> _____

THE ANTITRUST MOVEMENT

Big business caused problems for small businesses. A big business would lower its prices until small businesses, unable to offer the same low prices, went bankrupt. Consumers then had to pay higher prices because there was no longer any competition.

Americans demanded that Congress pass laws to control **monopolies** and trusts. Congress finally passed the **Sherman Antitrust Act**. However, the act was weak and did little to curb corporations.

> **Why did some people think trusts were bad for society?**
> _____
> _____
> _____

CHALLENGE ACTIVITY

Critical Thinking: Evaluate You are an adviser to the president. She has received complaints about big discount stores putting small, family-owned stores out of business by lowering prices. Make a list of advantages and disadvantages of large stores. Write a summary of your list and advise the president what to do. **HSS Analysis Skills HI 3**

Chapter 18 An Industrial Nation

Section 3

MAIN IDEAS

1. The desire to maximize profits and become more efficient led to poor working conditions.

2. Workers began to organize and demand improvements in working conditions and pay.

3. Labor strikes often turned violent and failed to accomplish their goals.

 HSS 8.12

Students analyze the transformation of the American economy and the changing social and political conditions in the United States in response to the Industrial Revolution.

Key Terms and People

Frederick W. Taylor author of *The Principles of Scientific Management*

Knights of Labor large labor union that included both skilled and unskilled workers

Terence V. Powderly Knights of Labor leader who made it the first national labor union in the United States

Samuel Gompers leader of the American Federation of Labor

American Federation of Labor group that organized individual national unions of skilled workers

collective bargaining workers acting together for better wages or working conditions

Mary Harris Jones union supporter who organized strikes and educated workers

Haymarket Riot a union protest in Chicago where strikers fought with police

Homestead Strike violent 1892 strike of Carnegie steelworkers ended by state militia

Pullman Strike strike of Pullman workers that ended in1894 when federal troops were sent to stop it

Section Summary

THE NEW WORKPLACE

During the Second Industrial Revolution, machines did more and more work. The unskilled workers who ran the machines could not complain about conditions, for they knew they could be replaced.

In the early 1880s **Frederick W. Taylor** wrote a book that took a scientific look at how businesses could increase profits. One way was to ignore workers and their needs. As a result, conditions for workers got worse.

> **What impact did Frederick Taylor's book have on America's workers?**
>
> _____
> _____
> _____

LABOR UNIONS

Workers began to form labor unions. The **Knights of Labor** started out as a secret organization. However, by the end of the 1870s, under the leadership of **Terence V. Powderly**, the Knights became a national labor union. The Knights included both skilled and unskilled members.

> Which union would have more power—a union of unskilled workers or a union of skilled workers?
> _____
> _____
> _____

The **American Federation of Labor**, under the leadership of **Samuel Gompers**, was different than the Knights of Labor. It organized national unions, and its members were all skilled workers.

Workers hoped that if they acted together—that is, if they used **collective bargaining**—they might actually be able to improve pay and working conditions.

> How did workers benefit from collective bargaining?
> _____
> _____
> _____

Many women participated in unions. **Mary Harris Jones**, for example, helped organize strikes and educate workers.

THE HAYMARKET RIOT

In 1886 thousands of Chicago union members went on strike. After police killed two strikers, workers met at Haymarket Square to protest the killings. Someone threw a bomb, and officers fired into the crowd. The **Haymarket Riot** ended with more than 100 people killed or wounded.

LABOR STRIKES

On June 29, 1892, at a Carnegie steel plant in Homestead, Pennsylvania, the **Homestead strike** began. Workers protested the introduction of new machinery and the loss of jobs. It ended in violence and death, and the union was defeated. Two years later, the **Pullman strike** over layoffs and pay cuts also ended in bloodshed. President Grover Cleveland sent federal troops to break the strike.

> Why do you think labor strikes often ended in violence?
> _____
> _____
> _____

CHALLENGE ACTIVITY

Critical Thinking: Write to Summarize Write a summary of the goals of the Knights of Labor. **HSS Analysis Skills HI 2**

Section 4

MAIN IDEAS
1. The late 1800s brought a wave of new immigrants from southern and eastern Europe and Mexico.
2. Some Americans opposed immigration and tried to enact restrictions against it.

 HSS 8.12
Students analyze the transformation of the American economy and the changing social and political conditions in the United States in response to the Industrial Revolution.

Key Terms and People

old immigrants immigrants who arrived in the United States before the 1880s and were mainly from western Europe

new immigrants immigrants who arrived in the United States in the 1880s or later and came mostly from eastern and southern Europe

steerage most immigrants traveled to America in this cramped and dangerous area on a ship's lower levels

benevolent societies neighborhood groups that offered aid to immigrants in poor areas

Chinese Exclusion Act 1882 law that banned Chinese immigration to the United States for 10 years

Immigration Restriction League a nativist group whose goal was to lower the number of immigrants

Academic Vocabulary

policy rule, course of action

Section Summary

NEW IMMIGRANTS

At the end of the 1800s, immigrants continued to arrive in the United States by the millions. These new arrivals were different from **old immigrants**. Old immigrants were mainly from western Europe. The **new immigrants** came mostly from southern and eastern Europe. They brought with them their own cultures and religions. Many "native" Americans blamed the new immigrants for crime, poverty, and unemployment and tried to stop immigration.

> What are the differences between old immigrants and new immigrants?
> _____
> _____
> _____

Getting to the United States was hard. Most
immigrants could only afford tickets in **steerage**.
When immigrants arrived they had to go through
immigration processing centers. In 1892 the federal
government opened a new processing center on
Ellis Island in New York Harbor. For the next 40
years, Ellis Island was the first American experience
for millions of immigrants.

> **What made immigrating to the United States so difficult?**
> _____
> _____
> _____

IMMIGRANT LIFE AND WORK

Most new immigrants settled in big cities. Usually
they moved into neighborhoods where others from
the same country lived. Few immigrants had any
money, and many lived in poorly built, overcrowded
tenements. Some received aid from **benevolent
societies**.

> **Why did immigrants live in poorly built, overcrowded buildings?**
> _____
> _____
> _____

Most new immigrants had to take low-paying,
unskilled jobs in factories, mills, or sweatshops.
Survival depended on the wages of all family members, so women and children also had to work.

Mexican immigrants who settled in the Southwest
worked as unskilled laborers. In addition, some
worked the large commercial farms found there.

OPPOSITION TO IMMIGRATION

Some Americans worried about immigration. Some
were prejudiced against immigrants. Others worried
that the new immigrants would drain America's
resources. Labor unions feared that industries
would hire immigrants instead of Americans.

In 1882 Congress passed the **Chinese Exclusion
Act**, which banned Chinese immigration. Later,
to appease the nativist **Immigration Restriction
League**, Congress also tried to make immigrants pass
literacy tests. President Cleveland vetoed that bill.

CHALLENGE ACTIVITY

Critical Thinking: Write to Defend You are an adviser to President
Cleveland. Write a paragraph explaining why he should veto the immigrant literacy bill. **HSS Analysis Skills HR 5**

Chapter 18 An Industrial Nation

MAIN IDEAS
1. New technology and ideas were developed to deal with the growth of urban areas.
2. The rapid growth of cities created a variety of urban problems.

 HSS 8.12
Students analyze the transformation of the American economy and the changing social and political conditions in the United States in response to the Industrial Revolution.

Key Terms and People

mass transit modes of public transportation, such as trolleys, subways, and commuter trains

suburbs residential neighborhoods outside of downtown areas

mass culture leisure and cultural activities shared by many people

department stores giant retail shops with large quantities of products

settlement houses neighborhood centers in poor urban areas

Hull House Chicago settlement house founded by Jane Addams and Ellen Gates Starr

Jane Addams co-founder of Chicago's Hull House

Section Summary

THE GROWTH OF URBAN AREAS

Immigrants accounted for part of the growth American cities experienced around 1900. Farmers replaced by machines also made their way to the cities. African Americans headed for cities with hopes of escaping the discrimination of the South.

The strain on urban space improved with the inventions of cheaper steel and the safety elevator. These two technologies made it possible to build skyscrapers. Cities could grow *up* instead of *out*. **Mass transit** systems, including elevated rails, subways, cable cars, and trolleys, helped ease traffic congestion of in the city. Mass transit, and the automobile, made it possible for Americans to move to the **suburbs**. People could live outside the downtown area and still work in the city.

The United States began to develop forms of **mass culture**. These are leisure and cultural

> **What are some of the reasons American cities grew around 1900?**
> _____
> _____
> _____

> **Why did people want to move to the suburbs?**
> _____
> _____
> _____

activities shared by many people. One reason these activities could be shared by more people were technological developments that made publishing cheaper. One effect of this was that the number of newspapers published in the United States soared.

The introduction of **department stores** also encouraged the development of mass culture. Mass culture included world's fairs, entertainment areas such as amusement parks, and open public spaces. Inexpensive train fares and entrance tickets allowed many people to enjoy public parks.

> **Why would department stores have an impact on mass culture?**
> _____
> _____
> _____

URBAN PROBLEMS

Despite many technological improvements to the nation's cities, tenement life continued to be a struggle. Often tenements had few windows through which air could flow. Many tenement buildings had no running water. These crowded and unsanitary conditions led to disease and health problems. Fire and crime also plagued urban life. City governments suffered from internal corruption and a lack of funds. This made it difficult to reform poor urban areas

Settlement houses, or neighborhood centers in poor areas, were a way for individuals to help immigrants. In 1889 **Jane Addams** and Ellen Gates Starr opened the **Hull House** in a poor part of Chicago. This would become one of the country's most famous settlement houses. Its staff helped neighborhood families. It also promoted child labor reform and an eight-hour workday.

> **Why were settlement houses built in poor urban areas?**
> _____
> _____
> _____

CHALLENGE ACTIVITY

Critical Thinking: Write to Defend Choose an example of mass culture you feel has had the greatest impact on Americans and write a paragraph explaining your choice. **HSS Analysis Skills HI 1**

Chapter 19 The Spirit of Reform

HISTORY–SOCIAL SCIENCE STANDARDS

HSS 8.12 Students analyze the transformation of the American economy and the changing social and political conditions in the United States in response to the Industrial Revolution.

HSS Analysis Skill HR 2 Students distinguish fact from opinion in historical narratives and stories.

HSS Analysis Skill HI 2 Students understand and distinguish cause, effect, sequence, and correlation in historical events, including the long- and short-term causal relations.

CHAPTER SUMMARY

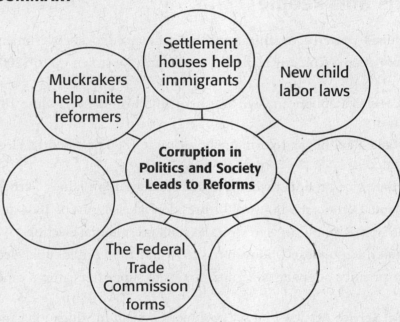

COMPREHENSION AND CRITICAL THINKING

Use the answers to the following questions to fill in the graphic organizer above.

1. Identify Look at the graphic organizer above. Identify four additional reform efforts made in the later 1800s and early 1900s.

_____

_____

2. Make Judgments Of the reforms you have identified for your web, which, in your opinion, was the most important? Write a sentence defending your answer.

_____

_____

Chapter 19 The Spirit of Reform

MAIN IDEAS

1. Political corruption was common during the Gilded Age.
2. Presidents during the Gilded Age confronted the issue of corruption.
3. In an effort to clean up political corruption, limits were put on the spoils system.

 HSS 8.12

Students analyze the transformation of the American economy and the changing social and political conditions in the United States in response to the Industrial Revolution.

Key Terms and People

political machines powerful organizations that influenced local governments

William Marcy Tweed Tammany Hall boss who may have stolen up to $200 million from New York City

Rutherford B. Hayes Republican Civil War hero who won the very close 1876 presidential election

James A. Garfield Republican reformer who won the 1880 presidential election and was assassinated

Chester A. Arthur vice president who became president upon James Garfield's death

Grover Cleveland Democrat who won the presidential elections of 1884 and 1892

Benjamin Harrison Republican who won the 1888 presidential election

William McKinley Republican who won the 1896 and 1900 presidential elections

spoils system practice of giving government jobs to supporters after a candidate wins an election

Pendleton Civil Service Act law that set up a merit system in which job candidates were chosen based on their abilities

Section Summary

CORRUPTION IN POLITICS

During the Gilded Age, local governments and the federal government were riddled with corruption. Boss-run **political machines** ran local politics. They would do anything to get their candidates elected. This included illegal activities such as buying either the votes or the vote-counters in an election. The bosses who ran the political machines also often stole for their personal gain. Tammany Hall boss **William Marcy Tweed**, for example, may have stolen up to $200 million from New York City for himself.

For what is William Marcy Tweed remembered?

_____

_____

In Washington, President Ulysses S. Grant's two terms both were branded corrupt. Many leaders in Congress who were involved in a railroad scandal also were not trusted by the public.

GILDED AGE PRESIDENTS

Republicans won two close elections in a row during the Gilded Age. In 1876 Republican **Rutherford B. Hayes** narrowly won the presidential election. In 1880 Republican **James A. Garfield** won another. Garfield's win was short-lived, however. He was killed and his vice president, **Chester A. Arthur**, became president.

In 1884 Democrat **Grover Cleveland**, who was known for his honesty, won the presidency. In 1888 Cleveland lost the electoral vote to Republican **Benjamin Harrison**. In 1892, however, Cleveland came back to win again and became president for a second term. In 1896 and 1900 Republican **William McKinley** served as president.

> Who served as president between Cleveland's two terms?
> _____
> _____

CIVIL SERVICE REFORM

Many Americans felt the civil service, or government jobs, was an area of corruption, and it was. Much of the civil service worked on the **spoils system**. Under this system, many civil service jobs were given to loyal party supporters rather than qualified candidates.

> What was the spoils system?
> _____
> _____

Presidents Hayes and Garfield tried to reform civil service. However, it was President Chester Arthur who was forced to respond to public demands for reform. He signed into law the **Pendleton Civil Service Act**. This act required many civil service applicants to take tests to prove they were qualified for the job.

> Circle the president that signed into law the Pendleton Civil Service Act.

> How did the Pendleton Act make sure that government jobs were filled by qualified people?
> _____
> _____

CHALLENGE ACTIVITY

Critical Thinking: Time line Make a time line of the presidents of the United States during the Gilded Age. **HSS Analysis Skills HI 1.**

Chapter 19 The Spirit of Reform

MAIN IDEAS

1. Progressives pushed for urban and social reforms to improve the quality of life.

2. Progressive reformers expanded the voting power of citizens and introduced reforms in local and state governments.

 HSS 8.12
Students analyze the transformation of the American economy and the changing social and political conditions in the United States in response to the Industrial Revolution.

Key Terms and People

progressives reformers who wanted to solve the problems of a fast-growing society

muckrakers journalists who exposed the corruption, scandal, and filth of society

John Dewey supporter of early childhood education

Joseph McCormack reorganized the American Medical Association in 1901 to bring together local medical organizations

direct primary allowed voters to choose candidates for public office directly

Seventeenth Amendment allowed Americans to vote directly for U.S. senators

recall a vote to remove an official from office before the end of his or her term

initiative procedure allowing voters to propose a new law

referendum procedure permitting voters to approve or reject a law

Robert M. La Follette Wisconsin governor who challenged the power of party bosses

Wisconsin Idea reforms put into effect by Wisconsin governor Robert M. La Follette

Academic Vocabulary

motive a reason for doing something

Section Summary

PROGRESSIVES WORK TO IMPROVE SOCIETY

Progressives were reformers. They wanted to solve problems caused by the fast urban growth of the late 1800s. Different progressives supported different social causes. But progressives all agreed that the federal government had to help fix the problems.

Journalists also became involved in reform. Many wrote articles about child labor, racial discrimination, slum housing, and other problems. Called **muckrakers** for the scandals they exposed, these writers encouraged voters to call for reforms.

> **Where did muckrakers get their nickname?**
> _____
> _____

Many progressives worked to reform the nation's overcrowded, dirty cities. They encouraged the passage of building laws. They also started settlement houses like Hull House. In addition, city planners and civil engineers helped make cities safer, cleaner, and easier to travel through.

> **How would new building laws affect the safety of tenement dwellers?**
> _____
> _____
> _____

Many progressive leaders focused on education. States began to pass laws requiring children to attend school. In addition, kindergartens opened to teach children basic skills. **John Dewey** was a champion of early childhood education. His motive was to help students learn problem-solving skills.

> **How did reformers make sure children got an education?**
> _____
> _____
> _____

Education reform also affected adults, particularly medical professionals. Under **Joseph McCormack**, the American Medical Association was reorganized in 1901. It made sure doctors nationwide were informed of changes and newly discovered knowledge in the medical industry.

EXPANDING DEMOCRACY

Reformers believed the voter should have more power. They favored the **direct primary**, in which voters choose candidates directly. They also favored the **Seventeenth Amendment**, which allowed voters to directly vote for U.S. senators. Another reform allowed **recall** votes to remove an official from office before the end of his or her term. Some states allowed voters to propose laws through **initiatives**. Some let **referendums** overrule laws.

In Wisconsin, Republican **Robert M. La Follette** challenged the power of the political machines. After he was elected governor in 1900, he began a series of reforms to reduce the power of the party bosses. This program, called the **Wisconsin Idea**, was soon used as a model for reform in other states.

> **What was the Wisconsin Idea?**
> _____
> _____
> _____

CHALLENGE ACTIVITY

Critical Thinking: Write to Compare Before the Seventeenth Amendment was ratified, how were U.S. senators chosen? Write an essay explaining both methods of electing senators, and which method you think is best.
HSS Analysis Skills CS 2.

Chapter 19 The Spirit of Reform

MAIN IDEAS

1. Reformers attempted to improve conditions for child laborers.

2. Unions and reformers took steps to improve safety in the workplace and working hours.

 HSS 8.12
Students analyze the transformation of the American economy and the changing social and political conditions in the United States in response to the Industrial Revolution.

Key Terms and People

Florence Kelley reformer who led the progressive fight against child labor

Triangle Shirtwaist Fire tragic fire at the Triangle Shirtwaist Company, in which 146 workers died because exit doors were locked, that led to improved factory safety standards

workers' compensation laws laws that guarantee a portion of lost wages to workers injured on the job

capitalism system in which the private businesses run most industries and competition determines how much goods cost and how much workers are paid

socialism system in which government owns and operates a country's industry

William "Big Bill" Haywood union leader of the Industrial Workers of the World

Industrial Workers of the World labor union founded in 1905 on socialist beliefs

Section Summary

CHILD LABOR REFORM

Another area on which reformers focused was the area of child labor. Children worked in a variety of jobs. Some sold newspapers. Some took care of boarders at their homes. Some did piecework. Many worked in industry. Sometimes children as young as five were sent to work in factories and mills.

Florence Kelley was a reformer who got involved in child labor. She was a board member of the National Consumers' League, which spoke out on labor issues involving women and children.

In 1912 Massachusetts became the first state to pass a minimum wage law. Other states soon followed. Congress, too, tried to pass laws protecting working children. However, the Supreme Court ruled many of these laws unconstitutional.

> How do you think a minimum wage law helped to protect children?
>
> _____
> _____
> _____

REFORM IN THE WORKPLACE

Progressives also worked to improve the work life of adults as well as children. For example, they favored an eight-hour workday and higher wages.

> How would an eight-hour work day change workers' lives?
> _____
> _____
> _____

Workplace safety also was a great concern. However, it took several terrible accidents for states and businesses to act. One of the worst was the **Triangle Shirtwaist Fire** in New York City. When the factory caught fire, many workers died because management had locked the exit doors.

Reformers also concentrated on getting **workers' compensation laws** passed. These laws would compensate, or pay, workers injured on the job.

> What is the purpose of workers' compensation laws?

Some industries complained that workplace regulations interfered with business. Often the courts agreed, using the Fourteenth Amendment to support their decisions. In *Lochner* v. *New York*, the Supreme Court ruled that states could not restrict the types of labor agreements between employers and employees.

Labor unions also tried to improve working conditions. Some wanted to take drastic measures, such as changing the economic system of the country to socialism. In **capitalism**, private businesses run most industries and competition determines how much goods cost and how much workers are paid. Under **socialism**, the government owns all the means of production. A new union based on socialist ideas was founded in 1905. It was called the **Industrial Workers of the World**. Under its leader, **William "Big Bill" Haywood**, it hoped to overthrow capitalism. It was not successful and was almost dead by 1920.

> How was the Industrial Workers of the World different than other labor unions?
> _____
> _____
> _____

CHALLENGE ACTIVITY

Critical Thinking: Write to Make Judgments Find a copy of the Fourteenth Amendment to the Constitution. Write a paragraph telling whether or not you agree with the courts' views of the Fourteenth Amendment and business regulation. **HSS Analysis Skills CS 4.**

Chapter 19 The Spirit of Reform

MAIN IDEAS

1. Female progressives fought for temperance and the right to vote.

2. African American reformers challenged discrimination and called for equality.

3. Progressive reform did not benefit all minorities.

 HSS 8.12
Students analyze the transformation of the American economy and the changing social and political conditions in the United States in response to the Industrial Revolution.

Key Terms and People

Woman's Christian Temperance Union group that supported abstaining from alcohol

Eighteenth Amendment amendment banning production and sale of alcoholic drinks

National American Woman Suffrage Association group that worked for women's voting rights, founded by Elizabeth Cady Stanton and Susan B. Anthony

Alice Paul founded a woman suffrage group that became the National Woman's Party

National Woman's Party woman suffrage group founded by reformer Alice Paul

Nineteenth Amendment a constitutional amendment giving women the right to vote

Booker T. Washington African American educator who encouraged other African Americans to improve their own lives rather than fight discrimination

Ida B. Wells African American journalist who publicized lynchings in her newspaper

W. E. B. Du Bois African American reformer who publicized cases of racial prejudice

National Association for the Advancement of Colored People organization that brought attention to racial inequality

Section Summary

TEMPERANCE AND WOMEN'S SUFFRAGE

In the late 1800s, new educational opportunities opened up for women. However, many male-dominated fields were still closed to women. So women's reform groups focused on two issues. One was the right to vote. The other was temperance, for many believed that alcohol caused society's problems. In 1874 the **Woman's Christian Temperance Union** was formed. This group fought for passage of local and state laws restricting alcohol sales. Their efforts were rewarded in 1919 when the **Eighteenth**

Why did many women want temperance?

_____
_____

What was the purpose of the Woman's Christian Temperance Union?

_____
_____

166

Amendment banned the production and sale of alcoholic drinks.

Meanwhile women struggled for the vote. By 1890 some western states had given women suffrage. In that year, the **National American Woman Suffrage Association** was founded. Reformer Carrie Chapman Catt became its president in 1900. **Alice Paul** founded another group that would become the **National Woman's Party**. It used publicity to draw attention to the issue of suffrage. Paul and others even went to jail. Finally in 1920, the **Nineteenth Amendment** to the Constitution gave women the right to vote.

Why do you think western states gave women suffrage before eastern states did?

_____

_____

_____

AFRICAN AMERICANS FIGHT FOR CHANGE

African Americans had won their freedom, but they still faced discrimination and segregation. African American educator **Booker T. Washington** urged blacks to focus on education and economic well-being rather than fight discrimination. **Ida B. Wells** focused instead on ending discrimination. She wrote about lynchings of African American men in her newspaper, *Free Speech*. **W. E. B. Du Bois** also used publicity. He and others founded the **National Association for the Advancement of Colored People** in 1909 to publicize the plight of blacks.

If you were an African American woman like Ida B. Wells, would you focus on women's suffrage or African American discrimination?

_____

_____

_____

THE FAILURES OF PROGRESSIVES

Chinese and Mexican immigrants and American Indians were left out by many reforms. Sometimes the reformers' help was not wanted. In 1911 the Society of Native Americans was founded to fight Indian poverty. But some Native Americans felt that adopting the ways of whites was destroying Native American heritage.

Underline the names of all the reformers identified in this section.

CHALLENGE ACTIVITY

Critical Thinking: Research to Discover Alcohol is legal in the United States today. Research what happened to the Eighteenth Amendment. Write a paragraph about your findings. **HSS Analysis Skills CS 1.**

Chapter 19 The Spirit of Reform

MAIN IDEAS

1. Theodore Roosevelt's progressive reforms tried to balance the interests of business, consumers, and laborers.

2. William Howard Taft angered progressives with his cautious reforms.

3. Woodrow Wilson enacted banking and antitrust reforms.

 HSS 8.12

Students analyze the transformation of the American economy and the changing social and political conditions in the United States in response to the Industrial Revolution.

Key Terms and People

Theodore Roosevelt vice president who became president upon McKinley's death

arbitration a formal process to settle disputes

Pure Food and Drug Act law stopping the manufacture, sale, or transportation of mislabeled or contaminated food and drugs sold in interstate commerce

conservation protection of nature and its resources

William Howard Taft president elected in 1908

Progressive Party political party, also known as the Bull Moose Party, which was formed so Theodore Roosevelt could run for president in 1912

Woodrow Wilson Democratic president who worked to regulate tariffs, banking, and businesses

Sixteenth Amendment amendment that allows the federal government to impose direct taxes on people's incomes

Federal Reserve Act law that regulated banking by creating Federal Reserve banks

Clayton Antitrust Act law that strengthened older federal laws against monopolies

Federal Trade Commission government agency that regulated business trade practices

Academic Vocabulary

various of many types

Section Summary

PRESIDENT THEODORE ROOSEVELT

Vice President **Theodore Roosevelt** became president when President McKinley was murdered. Roosevelt would be a very progressive president. President Roosevelt believed in a Square Deal for businesses, workers, and consumers. During a 1902 coal miner strike, Roosevelt forced mine managers to agree to settle the dispute through the formal process of **arbitration**. Roosevelt's Square Deal idea helped him win the 1904 presidential election.

President Roosevelt regulated some businesses. Muckrakers helped him. They focused public attention on industry problems. For example, one muckraker, Upton Sinclair, wrote a book called *The Jungle* on meat processing. The conditions he described led to a meat-inspection law. His book also led to the **Pure Food and Drug Act** of 1906.

Roosevelt also supported **conservation**. To help protect nature, 150 million acres of public land was saved from development under Roosevelt.

> Why do you think many mining, logging, and railroad companies opposed conservation?
> _____
> _____
> _____

TAFT ANGERS PROGRESSIVES

William Howard Taft became president in 1908. He felt Roosevelt had taken more power than the Constitution allowed, so he moved toward reform slowly. An angered Roosevelt ran against Taft on the ticket of a new party called the **Progressive Party**, or Bull Moose, Party. However, Democrat **Woodrow Wilson** won the election.

> Why did Theodore Roosevelt form a third party for the 1908 elections?
> _____
> _____

WILSON'S REFORMS

President Wilson immediately began to push for tariff reforms. These reforms also led to ratification of the **Sixteenth Amendment**, which allowed an income tax.

Wilson worked to regulate banking with the **Federal Reserve Act**. Passage of the **Clayton Antitrust Act** and the creation of the **Federal Trade Commission** both helped regulate big business.

> What was the purpose of the Federal Trade Commission?
> _____
> _____

CHALLENGE ACTIVITY

Critical Thinking: Write to See Connections Write a sentence or two explaining how muckrakers influenced the president of the United States. **HSS Analysis Skills HI 2.**

Chapter 20 America Becomes a World Power

HISTORY–SOCIAL SCIENCE STANDARDS

HSS 8.12 Students analyze the transformation of the American economy and the changing social and political conditions in the United States in response to the Industrial Revolution.

HSS Analysis Skill HI 3 Students explain the sources of historical continuity and how the combination of ideas and events explains the emergence of new patterns.

CHAPTER SUMMARY

Date	Event
1867	Secretary of State Seward purchases Alaska for the United States.
1867	
1898	The United States wins the Spanish-American War and puts Cuba, Guam, Puerto Rico, and the Philippines under U.S. control.
1899	
1900	Hawaii becomes a U.S. territory.
1903	Construction on the Panama Canal begins.

COMPREHENSION AND CRITICAL THINKING

Use the answers to the following questions to fill in the graphic organizer above.

1. Recall Which island system was the first to fall under U.S. control? Which was the last?

_____

_____

2. Make Inferences When the Spanish-American War ended, Cuba, Guam, Puerto Rico, and the Philippines all came under U.S. control. Under whose control were they before?

_____

_____

Chapter 20 America Becomes a World Power

MAIN IDEAS

1. The United States ended isolationism.
2. Due to its economic importance, Hawaii became a United States territory.
3. The United States sought trade with Japan and China.

 HSS 8.12
Students analyze the transformation of the American economy and the changing social and political conditions in the United States in response to the Industrial Revolution.

Key Terms and People

imperialism the practice of building an empire by founding colonies or conquering other nations

isolationism a policy of avoiding involvement in the affairs of other countries

William H. Seward secretary of state who arranged for the purchase of Alaska in 1867

subsidy bonus payment from the government to a private company

Liliuokalani Hawaiian queen who proposed a new constitution that gave power back to Hawaiians in 1893

consul general chief diplomat in a foreign nation

spheres of influence areas where foreign nations control trade and natural resources

John Hay secretary of state under William McKinley who announced the Open Door Policy

Open Door Policy policy stating that all nations should have equal access to trade in China

Boxer Rebellion a revolt led by a group of Chinese nationalists that was angered by foreign involvement in China

Academic Vocabulary

process a series of steps by which a task is accomplished

Section Summary

IMPERIALISM AND EXPANSION

The decades before 1900 were an era of **imperialism**. During those years, Europeans built great empires by taking control of other lands.

At first, the United States followed a policy of **isolationism**—that is, it avoided getting involved with other countries. That changed in 1867 when Secretary of State **William H. Seward** purchased

> Why do you think the United States favored isolationism before 1867?
>
> _____
>
> _____
>
> _____

Section 1, continued

Alaska from Russia. The United States then gained
the Midway Islands, part of Samoa, and Hawaii.

Sugar became a leading export in Hawaii.
Hawaii's economy collapsed when Congress passed
a **subsidy** that favored sugar farmers in the United
States. When Queen **Liliuokalani** proposed return-
ing Hawaii to the Hawaiians, the United States sent
Marines to take control of the islands.

> Why would a subsidy in the United States hurt the Hawaiian economy?
>
> _____
>
> _____

THE OPENING OF JAPAN

In 1854 the United States opened trade relations
with Japan. The first United States **consul general**,
Townsend Harris, negotiated a treaty that expanded
trade with Japan. In 1868 people who favored trade
came into power in Japan. They began a 40-year
period of modernization. In 1894 Japan defeated
China and became a major world power.

FOREIGN POWERS IN CHINA

After Japan's attack, other countries took advantage
of China's weakness to establish **spheres of influence**.
To make sure it could profit from trade with China,
U.S. Secretary of State **John Hay** announced the **Open
Door Policy** in 1899. This policy proclaimed that all
nations should have equal access to trade with China.

For many Chinese, resentment grew over the
amount of control foreign nations exerted over
China. In 1900 this resentment exploded in the
Boxer Rebellion. The Boxers attacked the walled
settlement in Beijing where foreigners lived.

> Why were the Boxers unhappy with foreigners in China?
>
> _____
>
> _____
>
> _____

For two months the Boxers laid seige to the settlement. Then military forces from several nations
arrived and defeated the Boxers. China was forced
to pay $333 million to various nations.

CHALLENGE ACTIVITY

Critical Thinking: Write to Evaluate On a map, locate the Midway
Islands, the islands of Samoa, and the Hawaiian Islands. Write a para-
graph evaluating why the United States wanted to control these islands.
HSS Analysis Skills CS 3.

Chapter 20 America Becomes a World Power

MAIN IDEAS

1. Americans supported aiding Cuba in its struggle against Spain.

2. In 1898 the United States went to war with Spain in the Spanish-American War.

3. The United States gained territories in the Caribbean and Pacific due to the Spanish-American War.

 HSS 8.12
Students analyze the transformation of the American economy and the changing social and political conditions in the United States in response to the Industrial Revolution.

Key Terms and People

Joseph Pulitzer publisher of the *New York World* who used sensational stories to sell more papers

William Randolph Hearst publisher of the *New York Journal* who used sensational stories to sell more papers

yellow journalism technique that exaggerates and sensationalizes news stories

Teller Amendment war resolution amendment stating that the United States had no interest in taking control of Cuba

Emilio Aguinaldo leader of Filipino rebels against Spain

Theodore Roosevelt lieutenant colonel who organized the Rough Riders to fight Spain in the Caribbean

Anti-Imperialist League American organization that accused the United States of building a colonial empire

Platt Amendment amendment to Cuba's constitution that limited Cuba's rights and kept the United States involved in Cuban affairs

Section Summary

THE ROAD TO WAR

During the 1890s Cuba rebelled against Spain, which controlled the island. Americans were sympathetic to the Cubans. Two Americans who supported the rebellion were **Joseph Pulitzer** and **William Randolph Hearst**. Both men were powerful newspaper publishers. They used **yellow journalism** and published exaggerated stories about the rebellion. This technique sold more newspapers and increased American support for Cuba.

Why did publishers such as Pulitzer and Hearst use yellow journalism?

_____

_____

_____

WAR WITH SPAIN

Hearst published a letter in which a Spanish official called President McKinley weak. Many Americans felt anger toward Spain. Then in Cuba's Havana Harbor, the U.S. battleship *Maine* exploded. No one knew the cause, but many assumed it was Spain.

In response Congress passed a resolution declaring Cuba an independent country. The resolution also included the **Teller Amendment**, which stated that the United States would not take control of Cuba. Spain reacted by declaring war on the United States.

Part of this war was fought in the Pacific. The U.S. Navy destroyed Spain's Pacific fleet in the Spanish Philippines. Filipino rebels led by **Emilio Aguinaldo** helped U.S. troops take Manila.

Other U.S. troops focused on the Caribbean Sea. These troops included the Rough Riders, led by future President **Theodore Roosevelt**. After the United States won battles both on land (in Cuba and Puerto Rico) and at sea, Spain surrendered.

> How did William Randolph Hearst promote the war against Spain?
> _____
> _____
> _____

> Why did Filipinos help the United States fight the Spanish?
> _____
> _____
> _____

NEW TERRITORIES

The peace treaty with Spain put Cuba, Guam, Puerto Rico, and the Philippines under U.S. control. A group of Americans formed the **Anti-Imperialist League**. They were afraid the United States wanted to build an empire. Despite their work, the treaty passed.

In Cuba the United States set up a military government. In addition, the United States added the **Platt Amendment** to Cuba's new constitution. It allowed the United States to stay involved in Cuba's affairs.

The United States decided to keep the Philippines. Filipinos fought the United States for their independence and hundreds of thousands died. The United States also kept Puerto Rico.

> Why do you think many Americans were opposed to the United States controlling more land?
> _____
> _____
> _____

CHALLENGE ACTIVITY

Critical Thinking: Write to Compare Write a paragraph that compares the purpose of the Teller Amendment with the purpose of the Platt Amendment. **HSS Analysis Skills HI 1.**

Chapter 20 America Becomes a World Power

Section 3

MAIN IDEAS

1. The United States built the Panama Canal.

2. Theodore Roosevelt changed United States policy toward Latin America.

3. Presidents Taft and Wilson promoted United States interests in Latin America.

 HSS 8.12

Students analyze the transformation of the American economy and the changing social and political conditions in the United States in response to the Industrial Revolution.

Key Terms and People

Hay-Herrán Treaty treaty to build a Central American canal that was rejected by Colombia

Philippe Bunau-Varilla engineer who planned a revolt in Panama against Colombia, with the help of the United States

Hay–Bunau-Varilla Treaty 1903 treaty agreed upon by the new nation of Panama and the United States to build a canal

Panama Canal canal built in Panama that shortened the Atlantic-to-Pacific voyage

Roosevelt Corollary President Roosevelt's warning that nations in the Western Hemisphere should pay their debts and "behave"

dollar diplomacy President William Howard Taft's policy of influencing governments through economic intervention

Academic Vocabulary

role assigned behavior

Section Summary

THE PANAMA CANAL

When ships in the Pacific headed toward Cuba to fight the Spanish, they spent weeks going around the southern tip of South America. People began to talk about cutting a canal across Central America.

In 1850 the United States and Great Britain had agreed to build a canal. They never did. France tried in 1881, but quit six years and $300 million later. Then Theodore Roosevelt became president. He sent Secretary of State John Hay to negotiate a deal with Colombia to lease land for a canal. Hay

> Why were foreign nations interested in building a canal through Central America?
>
> _____
> _____
> _____

and Colombian minister Thomas Herrán reached
an agreement, called the **Hay-Herrán Treaty**.
Colombia decided to reject it.

Engineer **Philippe Bunau-Varilla** told the United
States that the province of Panama was going to break
away from Colombia. When Panama revolted on
November 3, 1903, a U.S. warship kept Colombian ships
from reaching Panama to stop the rebellion. Panama
declared itself independent. Then the United States and
Panama agreed to build a canal with the **Hay–Bunau-
Varilla Treaty**. The **Panama Canal** opened about 10
years later.

> Why do you think Colombia was opposed to the United States building a canal?
> _____
> _____
> _____

ROOSEVELT AND LATIN AMERICA

In 1823 President Monroe warned European
nations to stay out of the Western Hemisphere.
The Monroe Doctrine, as this warning was called,
became a major part of U.S. foreign policy.

Roosevelt wanted the United States to maintain
control of the Western Hemisphere. By the time
Roosevelt became president, several Latin American
nations were in debt to European investors.
Roosevelt warned South American nations that if
they did not pay their debts, the United States would
step in. This is called the **Roosevelt Corollary** to the
Monroe Doctrine.

> How is the Roosevelt Corollary related to the Monroe Doctrine?
> _____
> _____
> _____

TAFT AND WILSON PROMOTE U.S. INTERESTS

President William Howard Taft tried **dollar diplomacy**. He wanted to control Latin America by
increasing U.S. business interests there.

President Woodrow Wilson ended dollar diplomacy. He did not like the role of big business in foreign affairs. He sent troops to stop unrest in nations
such as Haiti and the Dominican Republic.

> What is an advantage of dol-lar diplomacy over military intervention?
> _____
> _____

CHALLENGE ACTIVITY

Critical Thinking: Write to Sequence Write a paragraph that relates the
attempts U.S. presidents made to control Latin America.
HSS Analysis Skills CS 1.

Chapter 20 America Becomes a World Power

Section 4

MAIN IDEAS

1. In 1910 Mexicans revolted against their government.

2. The Mexican Revolution threatened United States interests economically and politically.

 HSS 8.12

Students analyze the transformation of the American economy and the changing social and political conditions in the United States in response to the Industrial Revolution.

Key Terms and People

Porfirio Díaz president of Mexico from 1877 to 1880 and from 1884 to 1911, a total of 30 years

Francisco Madero democratic reformer who began the Mexican Revolution

Mexican Revolution rebellion against Mexican president Porfirio Díaz

Victoriano Huerta general who took power in Mexico and had Madero killed

Venustiano Carranza led a successful revolt against Huerta and set up a new government in 1914

Francisco "Pancho" Villa northern rebel who continued to fight even after Huerta was forced out of office

Emiliano Zapata southern rebel who continued to fight even after Huerta was forced out of office

ABC Powers Argentina, Brazil, and Chile

John J. Pershing American general who pursued Pancho Villa through Mexico but never caught him

Section Summary

THE MEXICAN REVOLUTION

Porfirio Díaz was president of Mexico for 30 years. He was a harsh ruler. He punished his enemies and rewarded his friends. The United States invested more than $1 billion in Mexico during Díaz's reign. However, most Mexicans remained landless and poor.

In 1910 democratic reformer **Francisco Madero** started the **Mexican Revolution**. Madero and his followers forced Díaz to resign. The violence of the revolution caused many Mexicans to migrate to the United States. Madero's victory was short-lived. In February 1913 Mexican General **Victoriano Huerta**

Why did Mexicans want to migrate to the United States in the early 1900s?

_____

_____

_____

grabbed power. Huerta had Madero killed. The violence continued when **Venustiano Carranza** began
a revolt against Huerta's rule. Other rebel leaders,
including **Francisco "Pancho" Villa** and **Emiliano
Zapata**, also rose up against Huerta.

WILSON'S REACTION

This violence upset U.S. businesspeople. They
were worried about their investments as Mexico's
economy continued to weaken. The violence also
angered President Woodrow Wilson. He requested
and received permission from Congress to use force
against Mexico.

In 1914 Wilson sent the U.S. Navy to seize the
Mexican city of Veracruz to keep Huerta from
receiving supplies. Many Mexicans were furious. War loomed. Although the **ABC Powers**—
Argentina, Brazil, and Chile—offered to try to
negotiate a peace, conflict continued.

> **Why would the ABC Powers offer to help resolve the problems between the United States and Mexico?**
> _____
> _____
> _____

In July 1914 Carranza forced Huerta to flee.
Carranza set up a new government. This change in
government did not end the violence. Pancho Villa
wanted to overthrow Carranza. To gather Mexican
support, he started attacking United States citizens
on both sides of the border. Although Wilson sent
General **John J. Pershing** and 15,000 soldiers after
Villa, they never caught him.

> **How did Pancho Villa think he could win Mexican support for his rebellion?**
> _____
> _____
> _____

In 1920 Carranza was murdered by his aide. After
that peace gradually came back to Mexico.

CHALLENGE ACTIVITY

Critical Thinking: Write to Make Judgments Use the resources in your
library to find out more about one of the Mexican leaders mentioned in
this section. Then write a paragraph telling whether you think that person would be a good leader for Mexico, and explain your reasons why
or why not. **HSS Analysis Skills HR 1.**